STUDIES IN IMPERIALISM

general editor John M. MacKenzie

When the 'Studies in Imperialism' series was founded more than twenty-five years ago, emphasis was laid upon the conviction that 'imperialism as a cultural phenomenon had as significant an effect on the dominant as on the subordinate societies'. With more than ninety books published, this remains the prime concern of the series. Cross-disciplinary work has indeed appeared covering the full spectrum of cultural phenomena, as well as examining aspects of gender and sex, frontiers and law, science and the environment, language and literature, migration and patriotic societies, and much else. Moreover, the series has always wished to present comparative work on European and American imperialism, and particularly welcomes the submission of books in these areas. The fascination with imperialism, in all its aspects, shows no sign of abating, and this series will continue to lead the way in encouraging the widest possible range of studies in the field. 'Studies in Imperialism' is fully organic in its development, always seeking to be at the cutting edge, responding to the latest interests of scholars and the needs of this ever-expanding area of scholarship.

Cultures and caricatures of British imperial aviation

MANCHESTER
1824
Manchester University Press

SELECTED TITLES AVAILABLE IN THE SERIES

AIR EMPIRE
British imperial civil aviation, 1919–39
Gordon Pirie

THE COLONISATION OF TIME
Ritual, routine and resistance in the British Empire
Giordano Nanni

OCEANIA UNDER STEAM
Sea transport and the cultures of colonialism, c.1870–1914
Frances Steel

FLAGSHIPS OF IMPERIALISM
The P&O company and the politics of empire from its origins to 1867
Freda Harcourt

FROM JACK TAR TO UNION JACK
Representing naval manhood in the British empire, 1870–1918
Mary A. Conley

Cultures and caricatures of British imperial aviation

PASSENGERS, PILOTS, PUBLICITY

Gordon Pirie

MANCHESTER
UNIVERSITY PRESS
Manchester and New York

distributed in the United States exclusively by
PALGRAVE MACMILLAN

Published by MANCHESTER UNIVERSITY PRESS
OXFORD ROAD, MANCHESTER M13 9NR, UK
and ROOM 400, 175 FIFTH AVENUE, NEW YORK, NY 10010, USA
www.manchesteruniversitypress.co.uk

Distributed in the United States exclusively by
PALGRAVE MACMILLAN, 175 FIFTH AVENUE, NEW YORK, NY 10010, USA

Distributed in Canada exclusively by
UBC PRESS, UNIVERSITY OF BRITISH COLUMBIA,
2029 WEST MALL, VANCOUVER, BC, CANADA V6T 1Z2

British Library Cataloguing-in-Publication Data
A catalogue record for this book is available from the British Library

Library of Congress Cataloging-in-Publication Data applied for

ISBN 978 0 7190 86823 hardback

First published 2012

Typeset in Trump Medieval
by Servis Filmsetting Ltd, Stockport, Cheshire
Printed in Great Britain
by TJ International Ltd, Padstow

CONTENTS

ACKNOWLEDGEMENTS

The material and intellectual debts incurred in researching and writing this book are substantially the same as those acknowledged in an earlier companion volume, *Air Empire* (2009), in the same 'Studies in Imperialism' series (Manchester University Press). The research done for the two books overlapped for many years. Once again, therefore, it is a pleasure to thank the archivists and librarians whose work and collections made the inquiry possible, and to acknowledge the organisations which support record keeping and knowledge making. Those I have used most recently are the British Library, Cambridge University Library, the National Archives at Kew, the Post Office Museum and Archive, the Royal Aeronautical Society Library and the British Airways Archive and Museum.

Gratitude is again due the series editor, Professor John MacKenzie, for wise counsel. Anonymous readers gave generous, careful, constructive and encouraging comments on the book proposal and final draft. William Kentridge graciously allowed me use of his superb poster sketch. Henrik Larsen and Daniel Kusrow kindly answered queries. John Illsley let me reproduce a rare photograph he first unearthed. Philip Stickler helped locate sources. Kevin Winge gave me two writing tips.

Funds from the National Research Foundation in South Africa covered the costs of photographic reproduction and licensing. Study leave from the University of the Western Cape created a chunk of time to acquire additional evidence and to re-structure, trim, focus, reinforce and polish draft chapters. Most research and writing, however, was intermittent. Yet years of unpressured time were crucial for sniffing out and digging in likely sources, and for stumbling on others. Researching without deadlines, and without circumscribed scope and approach, has been an enormous and rare privilege; I wish the same 'blue sky' liberty for scholars whose inquiries will supersede mine.

LIST OF ILLUSTRATIONS

GENERAL EDITOR'S INTRODUCTION

There was an iconic moment in the patriotic career of the 'Iron Lady', Margaret Thatcher, when she showed her displeasure with British Airways. BA had decided to abandon (at least partially) its long-standing tail-fin design consisting of a stylised section of the Union flag. The company felt that it would improve its image in two ways – first by making it seem more 'modern' and secondly by rendering it more 'international' – by commissioning colourful, abstract tail-fin designs in a variety of different forms. Thatcher, visiting a BA display stand, spotted a model aircraft with one of the new designs. She promptly opened up the celebrated handbag, extracted a handkerchief and draped it over the offending tail fin. Famous for her many ways of expressing her displeasure, this one was eye catching and dramatic. BA eventually abandoned its new approach and the section of the Union flag reappeared. I preferred the colourful abstract designs, even if, in commercial terms, they were not particularly memorable in displaying the 'brand'.

Needless to say, as Gordon Pirie amply shows in this book, Imperial Airways was never behind-hand in its patriotic advertising and displays in the 1930s – Margaret Thatcher would have had no complaints in that decade. This is indeed the second of Pirie's books on Empire flying. The first, the award-winning *Air Empire*, looked at the pioneering days, the time of air exploration and explorers (both men and women), of the tentative first steps in the establishment of Empire air routes, a time when the rhetoric of flying read like that of African exploration in the nineteenth century. Now he has turned his attention to the 1930s when Imperial Airways was establishing itself as a worldwide airline, crossing oceans and continents in networks that could be strikingly mapped, rather like the maps of the sea routes of imperial shipping companies or of telegraph connections. Shipping 'lanes', telegraph cables and now air routes all represented not only the tentacles of technological progress but also the very ideologies and practices of imperialism. After all, it has been one of the fantasies of empires throughout history that they produce freedom of travel, facilitated by the imposition of 'peace' as in the 'Pax Britannica' and by aspects of global government, as well as through the colonial distribution of airports and technical facilities.

But the book contains a great deal more than just the fantasy of imperial flying. It deals with a period when it was still possible to see the technology as somehow distinctively British, even if readily emulated (and sometimes preceded and surpassed) by others across the world, with the planes and everything else genuinely produced under the Union flag. There is also material here on private flying, on the social and cultural dimensions of international aerial transport, on the participation and experiences of women (even if the great majority of fliers, both crews and passengers, were still men), and of the role of faster travel in forwarding the political and economic aims of Empire, as well as developing the leisure industry. Moreover, Pirie, very importantly, examines the reactions of many indigenous peoples, notably Arabs, Indians and Africans, who began to encounter planes for the first time. They were visited by them, occasionally overawed by them, often unimpressed by them. They were also sometimes, as David Omissi demonstrated in his book *Air Power and Colonial Control* in this same series, to be bombed by them when the British considered it politically and militarily expedient to do so.

The sources used by Pirie are extensive and varied. This helps to give the book its rich texture. It also ensures that Pirie's work is hugely entertaining as well as academically significant. Moreover, there is a certain elegiac feel to Pirie's narrative and analysis. This was a sort-of early golden age before the whole enterprise was dissolved in the violence of the Second World War. Somehow the great era of the passenger liners, particularly on the trans-Atlantic routes, seems to occur in the Edwardian Age of the twentieth century only to be destroyed by the First World War. From that time onwards, style, social tone and the cultural resonance of shipping changed. So it was with air transport after the second great conflict of that century. Air transport, as well as obviously aerial fighting and bombing, played an important part in that war, but it brought to an end the uncomplicated patriotisms and supposed imperial certainties of the 1930s. There was perhaps to be a last efflorescence of 'romantic' flying with the remarkable 'flying boats' of the post-war years, but major shifts in the geo-political balance were to produce dramatic changes during the two decades following the Second World War.

Imperial and patriotic contexts became unsustainable. Competition sent Imperial Airways – and its twin successors (nationalised in 1946) British European Airways and British Overseas Airways Corporation – into something of a nose dive, if somewhat temporarily retrieved in their subsequent (1974) amalgamation into British Airways. Decolonisation produced a host of independent states in Asia, Africa, the Caribbean, the Pacific and elsewhere, each aspiring – sometimes impractically – to prove their national maturity with the possession of an airline. Peoples who had formerly, at least in the minds of imperial fliers, looked up in awe and wonder, soon took to the air in vast numbers themselves. Flying became commonplace, ubiquitous and somehow ordinary, even if the technology remained remarkable. Changing planes in Dubai airport in the middle of the night (something I have done all too frequently in recent years) more than demonstrates the apparent universality of flight and fliers in the modern world. Waiting with hundreds of fellow passengers to join the huge A 380 on the flight (for example) to Beijing is an experience which seems aeons away from the world of the imperial airways described here by Gordon Pirie.

John M. MacKenzie

1

Introduction

'Think imperially, act imperially, travel imperially'. So ran headline text in a 1931 print advertisement for Britain's designated Empire airline.[1] The prominent typography used for the airline's name, *Imperial Airways*, made clear that the message was about flying imperially. Appearing in the official magazine of the Air League of the British Empire, the advertisement was part of that organisation's effort to cultivate 'airmindedness'. Britain too looked to the newest communications technology to entrench the Empire by using, managing and serving it better, and justifying it afresh. The Empire's airline offered new levels of real and imagined connectivity, mobility and status.

In 1931 it was still only possible to fly to, from and in a few parts of the British Empire as a commercial airline passenger. The London–Paris and Italy–India section of *Imperial*'s air route west of India had been open for two years; more or less regular air service to the Cape and Australia started in 1932 and 1934 respectively. Beforehand, several private flights had been made across the Empire on routes surveyed and first flown in late 1919 and early 1920. The imperial dreams behind those projects and flights, and the practical difficulties of bringing the Empire airways and commercial air services into operation, are the subject of *Air Empire*, a previous volume in this book series.

In this book, *Cultures and Caricatures of British Imperial Aviation*, the focus switches from the impetus for Empire aviation, and its geopolitical, commercial and technical frameworks, to the way airborne mobility itself expressed imperialism. Whereas *Air Empire* weighted events and discourse in the 1920s, here the emphasis is on the decade 1928–38, when the bulk of Empire civil flying occurred. However much of a struggle it was to activate and sustain British imperial presence and superiority in civilian air space, imperial qualities and characteristics lodged in the imaginations and behaviours of passengers and pilots, and in the eyes and minds of spectators and commentators. Eighty years later, the culture of flying imperially has become iconic, a caricature of late Empire.

Imperial flying was not just about machines, timetables and routes; it was also about ideas, values and practices. Imperial attitudes and actions were evident among those who flew privately along Empire air routes, and among airline passengers and crew. Their flights took them to and from London and Empire outposts as far as Cape Town and Sydney, and to and from places in between such as Nairobi and Calcutta. Imperial conduct, rank, style, discourse and (in)sensitivity were apparent on air journeys, in texts about flying overseas and in airline advertising images. Anything different would have been surprising, and perhaps vigorously countered had flying threatened to subvert imperial convention. A lot of that convention was about the supposed superiority of British and other white-skinned people.[2]

Imperial Airways, Britain's only overseas airline in the 1930s, served the Empire and articulated imperial benefits and standing. Its profile and performance were bound up with Britain's rating on the world stage, and with imperial propaganda. But the airline was not alone in the sky or on the ground. In the 1920s, the Royal Air Force was a conspicuous British representative abroad, its air bases strong evidence of a presence, and its formation flights (and bombs) an unmistakable sign of order and power. In addition, several individual pilots made remarkable flights across the British Empire. The first of these, in 1919–20, scattered the seeds of Empire aviation. After initial air crossings of the Atlantic (Alcock and Brown), and flights between England and Australia (the Smith brothers), and England and South Africa (van Ryneveld and Brand), other pioneering flights were made by several individuals and teams. Some were dispatched on survey missions as agents of Empire and airline.

The first flights (albeit only one-way) between Britain and outlying Empire territories were astonishing personal accomplishments and were statements of the individualism, courage and heroism which characterised early flying wherever it occurred. British pilots were not the only ones to make private long-distance flights between the two world wars, but their accomplishments were seized on as signs of national strength. Even after Alan Cobham's four path-breaking flights outward from London and back again between 1924 and 1927, individual pilots and their small teams of assistants continued trying to set height, speed or endurance records, or to notch up some other achievement. These aviators drew attention to themselves and to their country's modernity and industrial prowess. Successful pilots became national emblems. Impressionable youngsters looked up to them as role models; their triumphs triggered public celebration.

Some of the greatest public excitement in Empire aviation in the 1930s galvanised around private flights that showed off the Empire, its technological capability and the character of its citizens. Photographers focused on the celebrities and their aircraft. Commercial newsreels fed images to vast audiences around Britain and the Empire. After Cobham, Amy Johnson's feat of flying solo to Australia in 1930 gripped the public. The first-ever flight over Mount Everest in 1933, and the 1934 air race to Australia, further fired imagination. Like other flights which were made by individuals not travelling as airline passengers, these were not merely accomplishments to do with the reliability of engines and the learned ability to handle machine assemblages and to navigate. Flying imperially acquired its significance partly because it featured British people and planes. In varying degrees, private air journeys across the Empire showed off Britain. They unintentionally displayed and reflected imperial attitudes but were also aggressively manipulated as imperialistic propaganda. They also rocked and mocked the social and aeronautical establishment.

Not all men and women who flew over the Empire thought of themselves as its agents, and not all of them were adopted as imperial figures. Not all private flying was motivated by imperial display. Some pilots flew to win honours for themselves. Several pilots flew for pleasure, touring aerially on their own or as a pair. Their flights in light aircraft grew out of a long tradition of individual adventure, discovery and exploration. The self-recorded experiences of these air pioneers, and their encounters with people abroad, were cameos of imperialism. At best there was sleek service and repair, and there were sociable receptions. Unexpected landings generated fear, but also dependence and gratitude. Curious spectators could be a nuisance and a threat, and were dismissed in disparaging terms.

A great deal of the public fascination with flight revolved around routine commercial aviation on show at London's principal commercial airfield in the first half of the 1930s. But Croydon airfield did not monopolise interest or activity: some private flying was based at other airports. And, in the second half of the 1930s, *Imperial*'s flying boat base at Southampton became the hub of Empire aviation. The airline, however, did dominate Empire aviation, and its operations signified modern imperialism. Aircraft landing and departing with passengers from the furthest corners of the Empire could still tap a wellspring of pride, and create a thrill. Air freighting also registered imperial modernity and connectivity.[3] Flight was functional but it was also a spectacle. Destination marker boards in the airport building showed off the Empire; flying became an imperial performance. The airline itself worked hard to generate excitement about its services and to spark interest in the important people whom it made more mobile and who entrusted it with their lives. Expatriates ('birds of passage', indeed) were among these. Before 1939 few Empire residents who were not born British became airborne.

Whether in light private aircraft or on a commercial air service, flying to, from and in the Empire was expensive in the 1930s. Air travellers generally had private means or had their tickets paid by their state or corporate employers. The influential and elite of Britain and Empire revelled in the heavenly perspectives which flying offered, and its sensations of power, speed and efficiency. Height and might were aligned on a personal level. There was physical elevation to enjoy, as well as the rare cachet of air traveller. Encounters with other people in and from other lands were fleeting but interesting; disrupted flights offered adventure.

Paradoxically, flying also elevated air passengers when they were on the ground. The effect was most pronounced away from Britain where, even if air travellers did not personally feel superior to others, flying could easily accentuate the sense of being a race apart. Flying also made the inside of Empire more visible. It exposed the extent and variety of Empire in a new way. Hopping from one landing field to another unrolled a new perspective on Britain's territorial possessions, giving a first-hand impression of scale better than any map. Conversely, British residents or expatriates descending regularly from the sky at remote, inland ports of call gave the Empire a visible and audible presence in places where White people were scarce and were only recent settlers.

Flying imperially peeled back the living past. Empire air journeys brought strange sights and customs to the window. Flying across Africa and Asia was the inter-war equivalent of the European grand tour. Unlike a sea journey, a flight (generally below cloud level in unpressurised aircraft) sprung sights continuously. It was an encounter with past civilisations and with the cornucopia of contemporary cultures. It was a way of safely dipping one's toe into something foreign. Myths could be relived. Foreignness unravelled, and in the process the essence of Britishness came into sharper focus. The experience was recorded conscientiously in newspaper reports, magazine articles and travel diaries. What was the Empire if it couldn't be scripted?

Long before there were aeroplanes, British merchant ships flew the flag around the coastal margins of the Empire. They announced the presence of the metropolitan power, bringing and taking its managers, soldiers, settlers and traded commodities. Ocean-going liners and cargo ships provided physical links with Empire but also symbolised Britain's power and demonstrated its geographical reach. Their crews represented Britain abroad, and they took pride in keeping their Clyde-

or Thames-built vessels ship-shape as national emblems. Aura, order and national self-esteem were on show. Naming the prestige liners after queens and castles was part of British iconography.

Linking the British Empire by air presented the opportunity to reinvent these maritime signifiers. Furthermore, whereas there was never a single imperial shipping line, *Imperial Airways* could be used to represent both Britain and the Empire. The airline's home base at Croydon was the new concourse for priority, modern imperial exchange. Its workshops attended to the new engines of Empire. It was the launch pad for a new imperial project and its projectiles. The Empire airline displayed the British aeronautical industry and the nation's organisational ability. Air crews were roving ambassadors; ground crews were stationary representatives. Their behaviour and dress were meant to personify Britishness. Like their aircraft names, they were invested with qualities of reliability, courage and honour. Lapses from the stereotype of Englishman abroad did occur and were a matter of distress and denial. Matters of class and colour were raked up as appropriate.

Imperial Airways projected an idealised Britain to the Empire, and interpreted and refracted the Empire to Britons. Flying was 'made' by emotions, expectations, experiences, text and imagery. The airline's publicity pamphlets and press advertising were intended to propagate traffic. They used and reinforced archetypal images of the Empire as wilderness, hunting ground, playground and history lesson. Intended to promote the Empire airline and redress its London-centredness, *Imperial*'s touring exhibitions and commissioned documentary films caricatured Empire still further. Like the BBC Empire service which commenced in 1932, the Imperial Institute, the Empire Marketing Board (1926–33), the Colonial Empire Marketing Board (1937–39) and the Empire Film Unit, *Imperial Airways* was a source of Empire information and images which hoped to mould populist imperialism.

Airline advertisements, and magazine articles in the aeronautical press and in general interest serials, also helped to frame, reinforce and project imperial attitudes and behaviours. Accounts of Empire air journeys invariably worked with imperial tropes to reinforce the sense of social difference between imperial heartland and the colonial periphery. Book publishing added weight and deepened the divide by drawing attention to the wildness of Africa, the desolation of the Middle East, the strangeness of India and Asia, and the 'otherness' of people who lived there. The landing grounds used by British aircraft were safe islands in a foreign world.

Writing about and publicising imperial flying in the 1920s and 1930s created the first caricatures of Empire aviation. Words and images about long-distance air journeys, aircraft, landing grounds, passengers, crew and landscapes were necessarily selective and partial. Few pilots, whether working, adventuring or racing, had time or inclination to record and reflect on their activities. And, even if every airline passenger had travelled with a camera and a notebook, their impressions would have been mediated by purpose, taste and skill. Airline annual reports told only part of a much bigger story of aerial Empire. Advertising air travel laid down yet more simplification and generalisation. Posters and displays, and toys and postage stamps, sanitised and miniaturised Empire aviation.

The virtual mobility presented by 1930s texts and images, and enjoyed second hand by earthbound readers and viewers, was constructed in the moment by passionate proponents and critics, by jobbing journalists and paid publicists. The record is not necessarily accurate – or inevitably inaccurate – but its immediacy and spontaneity offer valuable glimpses of imperial feelings and attitudes. Imperial flight had a life beyond boardrooms, workshops and aeroplanes; it also lived in imagination and memory, in print and pictures, and in private and public desires. Furthermore, imperial flying harboured presumptions and prejudices. Eighty years later many of these are shameful in their arrogance, stereotyping and racism.

Leaving more traces of affection than fidelity, the reality and significance of flying imperially passed into memory after 1939. Thereafter, Empire flying became an avian specimen preserved in museums, archives and libraries, and cocooned in anecdotes, ephemera, illustrated histories and documentary films. Re-created in anniversary events, visual media and flight simulation computer software for hobbyists, imperial flying has braided into new celebratory forms. Accordingly, it has also become part of nostalgic conversations about 'the good old days' when Britain seemed great – or was great for some – and when flying was not yet a crushingly dull commodity. Flying imperially was different, and carried a different charge.

Notes

1 Reproduced in G. H. Pirie, *Air Empire: British Imperial Civil Aviation, 1919–39* (Manchester, 2009), p. 139.
2 P. B. Rich, *Race and Empire in British Politics* (Cambridge, 1986).
3 For a preliminary analysis see G. H. Pirie, 'British Empire air cargo in the 1930s: volume, variety, value', Unpublished paper.

PART I

Private flying

2

Aerial adventure

In a sense, all mobility is adventure. Even journeys made to destinations with predictable rewards contain uncertainties along the way; interesting transit is the adventure. The sheer act of flying can, indeed, be exhilarating. That apart, passengers in commercial aircraft did have adventures in the early days of Empire flying, but they did not seek these out except in the mildest possible way as fleeting, organised overnight stops at foreign places. *Imperial Airways* sold leisure, not adventure. Pilots who flew competitively, or on reconnaissance or aircraft test flights, or delivered aircraft to new owners, also had unintended adventures, but their purpose was not first and foremost recreational flying.

Flying for pleasure and adventure over long distances in the British Empire was part of the individualisation of imperial travel. Less is known about this aspect of imperial flying than about commercial airline flying. And journeys by just a few celebrities dominate what is known about private flying: ordinary pilots making comparatively unremarkable flights struggled to leave their mark. Yet the nature and scale of this activity counterbalance the dominant view of private imperial flying as mostly concerned with breaking records. It is also evidence of an ongoing culture of discovery and exploration, and of the ambiguous desire to flee elements of society while being its beneficiary, and depending on it *in extremis*.

British-manufactured light aircraft played their part in this late imperial movement. Various versions of the two-seat de Havilland 'Moth' especially were used by many British flying schools, of which there were more than sixty-two by 1934.[1] As a result, Moths – the equivalent of sports cars – were chosen by many who purchased their own private aircraft and by many who ventured abroad. The price was approximately £650 (equivalent to a semi-detached suburban house).[2] Some of more than 1,000 Moths built were shipped to the colonies and dominions. The private buyer market in Britain was not large: a mere twenty per cent of England's population in the 1930s earned more than £250 annually, and twice that amount was necessary to enable savings in a family.[3]

The original Moth, an open-cockpit bi-plane, owed its development to 'a fairy tale white knight', a sporty young man who had inherited a family fortune and contracted de Havilland to build him an aeroplane.[4] Wealth, leisure and sport were synonymous with all flying, into the Empire not least. Few people could afford to purchase or even rent a Moth; there were probably fewer than 100 'Moth flights' across the Empire by private pilots in the inter-war era. Examples are flights from England to Kenya and back in 1928 to collect photographs of the Royal visit for the *Daily Mirror*, and from Kenya to England at the end of a flying demonstration trip in 1929. A private flight from Iraq to Kenya was for a three-month photography and hunting trip.[5] In their Moth, a Mr and Mrs Chalmers flew from England to Iraq and back on a three-week holiday in 1931. Typical of the young, moneyed and leisured aerial adventurers, Sir Robert Clayton (aged twenty-four) flew his Moth to Egypt in 1932 to take part in an official desert survey. After his death due to infection, his young widow flew herself out to Egypt.[6]

The capital and operating costs of flying were high. Prior to his twenty-two-day Moth flight from England to Australia in 1929, Francis Chichester (twenty-eight) spent £20 just telegraphing for permits to fly over Egypt, Iraq, Persia, India, the Straits Settlements, Portuguese Timor and the Dutch East Indies. In the end, the entire adventure cost him £3,000.[7] When William Forbes-Sempill (the forty-two-year-old engineer pilot, and a figure of note in British aeronautical circles) flew himself to Australia and back in 1934/35 to represent the Royal Empire Society and the London Chamber of Commerce at Melbourne's centenary celebrations, he estimated that insurance payments alone amounted to a quarter of the value of his light aircraft.[8]

Flying partners

Private flying in light aircraft became a sport in Britain in the 1920s and 1930s.[9] Promoted at flying clubs, it was not necessarily leisured. It was mostly domestic. The pilot training and occasional shows and races at flying clubs around Britain generally looked inward at cultivating 'airmindedness' and equipping men (and some women) with flying and navigational ability. Some weekender pilots may have ventured into European skies. Relatively few flew out into the British Empire, although flying clubs were established in Britain's colonies quite early on. Whether for moral support, company or security, many people embarking on private adventure flights did so with a partner. Indicatively, between 1919 and October 1934, half of fifty-eight non-commercial and non-survey flights attempted between Britain and Australia (forty-two southbound from Britain) involved two or more people. Sixteen of the thirty-four completed flights carried two or more people.[10]

In 1925, Capt. T. Stack, the paid instructor at the Lancashire Flying Club, and B. Leete, an ex-RAF member of the Club, each bought Moths (£1,200 total) as part of a plan to fly to Australia. By then, the longest flights from Britain had been to Zurich (Cobham) and Italy (Hinkler). The five-day 'sporting venture' in which two RAF officers hoped to fly home to England on leave from Karachi in March 1926 faced engine trouble before they had covered 200 miles. Stack and Leete had been RAF colleagues at Hinaidi, Iraq. Setting out from England in November 1926, their journey included trick flying during the Baghdad Races, an emergency landing in the Persian Gulf and an engine overhaul at Bushire. They reached Karachi, 5,541 miles away, after fifty-four days. The award of the Air Force Cross preceded the crash at Lahore (with Lady Hailey, wife of the Punjab Governor, aboard) which ended their Australia attempt.[11]

En route to Lahore, Stack and Leete landed in New Delhi just hours before the Secretary of State for Air (Sir Samuel Hoare) and his official party touched down on their historic flight from England to India.[12] Sir Sefton Brancker (the Director of Civil Aviation) followed independently as a member of the party, and found 'an all-red trail of admiration and enthusiasm' for the two pilots' sporting effort. No doubt he meant well by his sporting reference, but it did distinguish subtly between weighty Empire events and those conducted by members of a provincial flying club. When he opened discussion on a paper read to the Royal Aeronautical Society by Stack, Brancker squeezed out admiration by making the pilots of diminutive aeroplanes seem larger than life. Not only were British people everywhere delighted with this British enterprise, he said, but French, Italians, Egyptians, Palestinians, Iraqis, Persians and Indians were 'sincerely impressed by the performance of these little machines and by the personalities of their pilots'. In India special trains took crowds to see the exhibition flights given by Stack and Leete. Flying clubs were formed in the subcontinent in short succession.[13]

Adventure of sorts propelled Richard Shuttleworth's flight to India in January 1933. The twenty-four-year-old 'amateur motor engineer' left his family's private flying ground at Old Warden near Bedford to take part in the Viceroy's Cup horse race at Delhi. He was accompanied by an RAF flight lieutenant in a second aircraft.[14] Kenneth Gandar Dower (twenty-six), who epitomised England's leisured elite even better than Shuttleworth, had flown himself and his friend (Angus Irwin, RAF) to Madras the year previously. Lord Londonderry, then Secretary of State for Air, admired this amateur sport of long-distance fun flying which made it possible to avoid 'a humdrum life'.[15] The flight in a light plane named *Spirit of Fun* was not as 'supremely insignificant' as Dower thought: he did not intend to set any records but, by chance, he and Irwin were the first to fly from England to Madras (two previous attempts had failed). Welcoming them were, literally, two men (from Shell) and a dog.[16]

Youth, wealth, time and an interest in machines and speed were predispositions to flying. Two young motor-racing drivers flew themselves from England to compete in the 1934 South African Grand Prix and started their return journey the day after their successes.[17] In 1933, the Earl of Ronaldshay was flown to India on business in a light plane.[18] Leaving his wife in London, a Grenadier Guard, Mr Clifford, flew to Cape Town in a light aircraft with a male friend in September 1934.[19]

Men did not only fly with other men. Several women accompanied their husbands (or a male pilot friend or instructor) on long-distance private flights which were not made competitively. Like proto-feminist domestic recreational flying (at clubs, displays and races) joint male–female flights in and round the Empire were not headlines in newspapers or cinema newsreels. Mostly, they were a footnote in history. One is the 1931 flight Mrs F. K. Wilson (founder-proprietor of Kenya's Wilson Airways) took with a male pilot from Zanzibar to London via the Gambia; they spent eighty hours in the air.[20]

The Cobhams were one of the first married couples to fly into and across the British Empire on a private flight. In November 1927 Sir Alan Cobham set out from London on a 25,000 mile, six-month flight around Africa.[21] He and his wife, Lady Gladys, left two young boys in England for several months. The hazardous trip drew no adverse social comment in public. Accusations of neglectful mothering did not even surface when Lady Maïa Carberry, aged twenty-four, died at the controls of her Moth in Kenya in 1928. By then she had already been separated from her two-year-old daughter for months while travelling overseas with her husband, including a spell in London learning to fly.[22]

Flying just months after Lady Hoare had accompanied Sir Samuel on an *Imperial Airways* aircraft delivery flight to the Middle East, Lady Gladys recalled her round-Africa journey in a military service aircraft as 'seven of the most perfect months I have ever had in my life'. She slept in one of four bunks aboard, typed thousands of words and pondered how flying might relieve the loneliness of women in the outposts of Empire. She probably did help to keep up European appearances, not least by making the male crew (co-pilot, two engineers and photographer) more circumspect in their conduct with women. Her husband may have stopped feeding grapes seductively to other women at colonial picnics. There may even have been an end to playful filming of bare-breasted 'dusky beauties' carrying pails of water to wash Cobham's aircraft, and an end to inviting young African girls to pose in the aircraft cockpit. Indeed, such was the sexual decorum in Lady Gladys' presence that on the momentous occasion of crossing the equator, Cobham suggested he would do something out of the ordinary by kissing his wife in front of the aircrew. Lady Cobham retorted that it would be even more extraordinary if the whole crew kissed her in front of him.[23]

The Cobhams were not the only married couple to fly into the Empire. In 1928, Flt/Lt R. Bentley and his new wife completed a honeymoon flight from London to Cape Town and back in a Moth. When delivering a new aircraft to an air survey company in southern Africa in 1930, the manufacturing company's pilot took with him from London his wife and a mechanic. Mostly, however, couples flying overseas did so without an aeronautical purpose. In mid 1931, Mr and Mrs Humble (British residents who learned to fly while visiting South Africa and purchased a light aircraft there) flew themselves from Cape Town to England in eighteen days. Mr Humble, a director of a British engineering firm, spent ten days doing business along the way. A couple from Southern Rhodesia flew to London in 1938. The Australian route also beckoned. Two months into their marriage in 1935, Mr and Mrs Littlejohn set out in their private plane for his Australian home. In December 1936, Mr C. Kelman and his wife flew to their Australian base from England.[24]

The veteran British armed forces commander, Brig.-Gen. A. C. Lewin and his sister-in-law left Nairobi for England in 1933 but their aircraft was destroyed in a storm at Mongalla (Sudan).[25] Lewin, an accomplished pilot, seemed jinxed. Later, aged sixty-three, he and his wife departed by air from England for their Kenyan farm in 1937. They got most of the way before making a forced landing in the Sudan. They were rescued (see Chapter 6). Worse had happened. Five years earlier, a couple on their honeymoon trip from Nairobi to South Africa crashed in their light aeroplane flown by the groom, Capt. R. Ussher, a well-known Kenyan settler. An African stumbled across the wreckage and the flight log book in thick forest fifteen miles from Mbeya (Tanganyika). Also at the scene was the groom's suicide note that he wrote after failing to correct 'bad cloud spin', surviving the crash and then seeing his new wife dead.[26]

Small press notices record other pairs flying imperially, but seldom show whether women shared the flying and navigation. It is unlikely they were merely passive passengers. In 1928 Capt. W. Lancaster and Mrs (Jessie) Keith Miller flew together from London to Australia in a

Figure 1 Mrs Cleaver and Mr Drew with family, friends and Moth at Croydon before flying to India, 1929.

light plane; they were known to be having an affair.[27] Mrs Adelaide Cleaver (identified only as the daughter of the Northern Ireland Minister of Finance, H. M. Pollock) flew to India and back in a Moth, Mr D. Drew piloting, starting out from Croydon in March 1929. Over the final stage of her journey she made a scheduled *Imperial Airways* flight from Paris to London. Also in 1929, Flying Officer L. Murray and his wife flew in a Moth from Peshawar to Melbourne. Miss Helen Silver, a young woman from Kenya who had been visiting relatives in England, began her return home in a dual-control light aircraft with Capt. Cameron in 1930. Unlike this duo, when Miss Winifred Spooner and flying officer E. Edwards started out on their flight to the Cape they were intent on doing it within ten days. Their hopes – but not their lives – ended in the Mediterranean.[28]

The following year, 1931, Delphine Reynolds (the unmarried and independently minded daughter of a Liverpool M.P., Sir James Reynolds) flew south along the West African coast in a light aircraft. She and her flying instructor, Flt/Lt W. Pudney, surveyed rivers with a view to their use for a seaplane service.[29] Another allegedly nineteen-year-old was reported to have flown from London to Kenya in 1931. In fact, the New Zealander was thirty-two when she set off in October with Mr F. ('Bunny') Richards, a naval surgeon-general's son whom she met at a cocktail party in London. Aline Barton lied about her age because she felt a newspaper reporter showed no interest in women pilots aged between twenty and ninety. Over a relaxed month-long flight that Aline was making simply as a fun way of getting to Kenya, she piloted and Bunny navigated the course via Turkey and Syria. The partnership was a matter of convenience not romance.[30]

In a busy 'flying season' in 1931, the Hon. Mrs Westenra (Lady Bailey's sister, and sister-in-law of the wealthy South African mining magnate, Sir Abe Bailey – see below) set out for Cape Town via the East African air route in her Moth. Capt. R. Macintosh piloted. Considerably quicker than her sister's 1928 odyssey, the nine-week-long 23,000 mile holiday air tour took them back to England on a hazardous track across West Africa.[31] It was there that Mrs Keith Miller's aircraft was wrecked while she was on her way from London to Cape Town early in 1935.[32]

After a light-hearted challenge taken up at a post-party breakfast in London in November 1931, Mr R. Richards took off for Australia with Lady Chaytor in March 1932, getting as far as India.[33] Simultaneously, the American Margery Durant (the wealthy, three-times divorced forty-six-year-old daughter of the founder of General Motors) was flying from Egypt to Uganda in her privately owned amphibian aircraft. The south-bound course of the three-month return trip was described by Durant's pilot as one 'from Civilization to Barbarism'. It ended with a hunting safari in Uganda assisted by sixty porters who had previously served in the making of the Hollywood film 'Trader Horn'.[34]

Prompted perhaps by Durant's trip, in October 1932 the widow of a wealthy American resident in England left her Southampton home for a trip to the Near East in her luxurious flying boat: Mrs J. Jones was accompanied by her secretary, her son and two mechanics. Their pilot had tasted victory in the prestigious Schneider Cup air races.[35] Some flights never took off. Miss May Spinks (a Harrogate 'resident' who had visited Africa seven times, flown over sections and driven from Nairobi to Cape Town) proposed to fly leisurely from Australia to England in 1933 with the pilot who accompanied her for several months on an aerial tour of Central and Northern Australia in 1932, but the plan foundered.[36]

The British Empire was not the exclusive domain of ultra-long-distance British private flying. During a leisurely ten-month aerial tour in 1928/29 (flying only in the morning and only in fair weather), Vicomte de Silbour and his wife Violette (daughter of Gordon Selfridge, the department store owner) flew their Moth across North Africa and the Middle East to India, Burma, Indo China and Japan. From there they sailed the Pacific, flew across the United States and crossed the Atlantic by sea. The forces of Empire gathered to assist them in the Persian Gulf: 'fearing an Arab attack on a British adventurer, an RAF punitive escort was sent out'. It was a useful training exercise, and the couple must have appreciated watchfulness after making an emergency landing and taxiing for thirty-five miles.[37] In 1930, Mrs Victor Bruce, a rally car driver, long-distance motorist and endurance racer, flew a similar track in the northern hemisphere. She bought her light aircraft on impulse for £550 in a shop in Burlington Gardens, London. Maps from Stanfords on the Strand cost £50. Her solo flight followed the standard air route to Bangkok before veering north to Hanoi, Hong Kong, Seoul and Tokyo. Turning flight into a commodity, her plane was put on show at Embankment underground station for a month.[38]

Persia and Russia beckoned Mrs Edwin Montagu, widow of a former Secretary of State for India. In March 1931 she set out on a six-week flying tour in a Moth. One forced landing, one fire and one year later, the Russian authorities refused her and her private pilot permission to fly via Siberia to the Far East. Undeterred, the two set out for Baghdad and Rangoon.[39] Male company was probably appreciated by Lady Blanche Douglas, who made an emergency landing in Persia while flying in her own aircraft with Flt/Lt Ogden from England to India in 1934/35.[40]

Unusually, in 1932 two licensed women pilots left England without male company on a flying trip across Africa. Their somewhat haphazard flight began with a series of minor incidents in North Africa.[41] Miss Audrey Sale-Barker, a noted skier and an elegant fixture of the London debutante season (and its French satellites), was from a moderately prosperous family. She did not rely on financial support from her doctor and actress parents who lived in a smart new apartment block behind Sloane Square in London. Instead, she learned to fly, aged seventeen, after winning free lessons in a competition for flying aptitude at the 1929 Aero Exhibition at London's Olympia. Provided she took a chaperone, her parents allowed her to fly out on a Christmas visit to Cape Town to see their friends, the Governor-General of South Africa and his wife, Lord and Lady Clarendon. Having turned twenty-one and having spent an inheritance on a light plane, she teamed up with twenty-three-year-old Joan Page, daughter of Sir Arthur Page, the Chief Justice of Burma.

The 'girl pilots' never confirmed press reports of how they managed to avert the prohibition on women flying across the Sahara without a male escort, but they may have taken turns travelling overland and flying with a male passenger. They kept the press at bay until returning north from Cape Town when they crashed in the dark in foul weather after departing from Moshi in Tanganyika. Filling the vacuum of news, the papers in Britain became almost hysterical. The most that can be said after all the concocted and exaggerated reporting about lions and wild men is that the girls were badly injured and, after a grim night, they did give a Masai herder a message written in lipstick calling for help. The girls were spotted and rescued (one by air, the other by land), and were hospitalised in Nairobi.[42]

Worse was possible, as noted by the lone German pilot who landed in the Gambia after engine failure. Her account of flying to the Niger, and then to India, Australia and South America, appeared in English translation in 1935. In the remote African bush she resorted to an extreme show of strength to persuade a local chief to 'loan' her donkeys to fetch help. She claimed she was a French queen, and would summon troops to shoot his cattle and women. Her ruse failed. Finally she did as her African 'adviser' suggested, and slashed the arm of the chief with an axe blade.[43]

In the category pilot-cum-passenger, Mary du Caurroy, the Duchess of Bedford, had the highest inter-war profile. An accomplished and dedicated surgical nurse and radiologist in the private hospital she built on Woburn Estate, the sixty-three-year-old Duchess took to the air for not uncommon reasons. She sought solitude away from the landed aristocracy she found boring, from drawing-room life she hated and from a son she barely knew. She also wanted pleasure away from a cold marriage to the Duke, who threatened to close the hospital. Yachting, angling, skating and shooting had helped. Flying was an even more exhilarating hobby – and it eased her hearing problem.[44]

After several short flights round Europe (her maids and courier travelling overland), in 1928 the Duchess flew to India. She had grown up there as the daughter of a chaplain who became the Archdeacon of Lahore. It was there that she also met her future husband, equerry to the Viceroy. The aircraft was a British-engined, Dutch-built Fokker light aeroplane flown by two much younger men. On the way, the Duchess did some flying under the direction of her personal pilot, Capt. Charles Barnard. Courtesy of the Bushire Resident, and British hospitality from the Anglo-Persian Oil Company at Abadan (near Basra) and at Ispahan, they survived an excruciatingly hot and humid ten-week stay in the Persian mid-summer waiting for a replacement engine. The Duchess's adaptability and imperturbability were clear. The flight to India was intended to be in both directions but, after propeller trouble was diagnosed at Karachi, she returned to England by sea. When Barnard and his co-pilot eventually made a record-breaking flight back to England from India in four-and-a-half days they took with them a case of tea instead of the Duchess.[45]

Barnard flew the Duchess to India and back in eight days (ninety hours airborne) in August 1929. In 1930 they flew to South Africa and back, spending 175 hours in the air. It was not the adventure with two dashing men that caused controversy. Rather, it was the Dutch-built Fokker aircraft that went on display in the Strand. One member of the Air League's executive committee was reported to have resigned in protest. A patriotic voice retorted that the plane's engine, tyres, plugs, maps, nuts and bolts, compass, oil, petrol, and even fountain pen for the flight log, were British products. Nobody ventured an opinion about whether receiving assistance overseas from people other than the King's subjects was unpatriotic. Should the Duchess have scowled at Egyptians who stood round her aircraft holding palm fronds for beating away any insects that might clog the radiator and air intakes?[46] Not everyone was surly: Brancker welcomed the Duchess at Croydon, she dined at the Royal Geographical Society and the Savoy, and had to deal with the usual deluge of letters and invitations.[47] Barnard wasn't finished and, in December 1930, he took off again from Croydon. The plane he piloted was owned by the Fourth Earl of Lovelace (twenty-five), who wanted to get to his coffee shamba at Babati (Dodoma), Tanganyika. The two got as far as Tripoli before crashing.[48]

As individuals made more flights around the Empire chasing their dreams, the chances of setting a speed or distance record, or otherwise making a mark, began to slip away. Sponsors became more reticent. Newspapers watched for spectacular failures as much as successes. A completed flight was no guarantee of public acclaim. Even in the glut, however, being British, or being resident in the 'white' Empire, was the most fortunate and desirable status.

The first Indian pilot and mechanic to fly from India to England would have had good reason to feel disgruntled after landing early in 1930. Mr R. N. Chawla and his seventeen-year-old assistant, A. M. Irani (whose *nom de l'air* was 'Engineer'), were not fêted. Irani, the son of a rich Karachi merchant who provided the aircraft, may have been used to more attention. Chawla and Irani's flight attracted little interest in the London press or in aviation offices. Infamy was more in attendance. At the suggestion of India's Director of Civil Aviation, the Air Ministry arranged for a representative to meet the two Indian pilots at Croydon. In the event, the pilots lost their way from France and landed near Thetford, Norfolk. The following day they found Croydon. On hand to meet them was the man from the Ministry, the Mayor, representatives of the Secretary of State for India and the High Commissioner, and a small press pack. The Viceroy sent a telegram. The two men were invited to the India Office. The Air League did not arrange a banquet. Instead, Indian students in London arranged a reception. The Air League did rather better in its mouthpiece, carrying Chawla's own account of the flight and noting a reception from the Mayor of Croydon.[49]

To his credit, the acerbic C. G. Grey offered some publicity in his *Aeroplane* magazine. He managed not to be patronising, but he deflected attention away from aviation and stressed the 'considerable ethnological interest in Chawla's flight'. Grey was relieved by Chawla's Parsee identity (Irani's Parsee identity went unmentioned). Here, he said, was an Indian who was as much a foreigner in India as were the English or the Pathans. Chawla evaded Grey's distaste for Indians 'of Dravidian or other aboriginal / indigenous stock which supplies the Babus and other objectionable or disturbing elements in the population of India'. Whatever her views on Empire politics, Mrs Sempill's diplomatic skills were only marginally finer than Grey's. At a public meeting attended by Chawla and Irani she acknowledged the skill and endurance of Indian pilots, but added indelicately that 'one swallow did not make a summer'.[50]

Polite innuendo and inattention repressed any threat posed to the smug colonialist assertion and representation of British cultural and technological superiority. The triumph of Western scientific rationality that had reached a crescendo in aeronautics was undiminished. Chawla had learned to fly at the Nottingham Aero Club in 1928, and his aeroplane was neither invented nor built in India. The technical success of Indian pilots could be brushed aside as an aberration from the mysticism that some thought manacled the Orient. Nobody knows what Mrs Sempill might have said if she had found out about Chawla's flight from India to South Africa later in 1930. C. G. Grey wondered why Indians would want to fly to a place where they met such hostility. He was not referring to England. The Aga Khan, at least, was proud. He stumped up the prize for the first flight by an Indian to South Africa, and funded the £500 reward that Irani won for being the first Indian to fly solo between England and India. He snatched the prize from under the nose of J. R. D. Tata (son of the wealthy Indian industrialist and, at twenty-seven, not yet the father of Indian aviation). Flying solo from Karachi to Croydon, Tata crossed with Irani already flying home. Irani's first experiences of British hospitality would have been moderated by the assistance he received from the Anglo-Persian Oil Company at Abadan and from the local Governor when he made a forced landing there on his abortive flight to Cape Town in 1930.[51]

Flying solo

Partnered flying across the Empire attracted less attention in the 1920s and 1930s than did long-distance journeys flown solo. This was so even though the number of solo and partnered attempts and successes was not markedly different. But solitary determination and heroics seemed the greater. Abandoned flights were worthy but not spectacular. In the public eye they registered lower than crashes and disappearances, which caught attention for a brief, sad moment. Flights accomplished in record time claimed the headlines – successful 'slow' flying quickly became unremarkable. Accordingly, a handful of solo flights has come to dominate the story of Empire private flying. In the Britain–Australia case, for example, eleven of twenty-eight solo flights (from fifty-eight private flights) recorded in one listing up to May 1934 are footnotes in history: they were not completed (including two by pilots who had already flown one trip successfully). Of the seventeen completed flights, barely half were made by recognisable 'names'.[52]

Like the giant American figures of Charles Lindbergh and Amelia Earhart in aviation history, the British Empire has its own pantheon of famous solo pilots. The Australian Charles Kingsford-Smith is known for three flights between Britain and Australia and, aged thirty-one, for being the first person to cross the Pacific by air, flying from the United States to Australia. In 1930 Amy Johnson became the first woman to fly from Britain to Australia, an achievement eclipsing Francis Chichester's similar flight only months previously. She subsequently flew the trans-Africa route twice. Her husband, Jim Mollison, whom she married in 1932, made solo flights across Africa and along the Britain–Australia route. Bert Hinkler, Jean Batten, Charles W. A. Scott and Jim Broadbent are forever associated with solo flights along the Empire airway through the Middle East, India and South East Asia. All were part of the brief and rare age when aviators were named and photographed.

The African airway created fewer star pilots. Among the better known record breakers were Flt/Lt R. Bentley (who in 1927 halved van Ryneveld and Brand's time to the Cape), Lt Patrick Murdoch (who did the same again the next year) and Alex Henshaw.[53] Two solo pilots who did not break trans-Africa records, but who are arguably better known, are Lady Bailey and Lady Heath. Beryl Markham, who lived in Kenya as the daughter of an English settler from the age of three, was a more authentic 'African female soloist'. She emerged to prominence in the happy-go-lucky colony first as a racehorse trainer and then as a private pilot after succumbing to T. Campbell Black's 'romantic and imperial reverie' about flying being empowering and creating 'a feeling of possession that you couldn't have if you owned all of Africa'.[54]

Private flying grew vigorously in Kenya in the late 1920s. It was during the colony's first air fair in 1928 that one of Markham's contemporaries, Maïa Carberry, died in an air crash. The wife of a renegade

Irish peer who had settled on a coffee plantation in the colony, Maïa had recently completed the first non-stop flight from Mombasa to Nairobi (in her aeroplane named *Miss Propaganda*). Newly licensed, she left England a matter of months earlier explaining how owning a little 'air-car' could revolutionise settlers' lives by enabling morning shopping trips into Nairobi, by making it possible to attend horse races and local amusements, and for summoning a doctor. Referring also to her husband's two aircraft, Maïa added that 'we shall use our air fleet for the benefit of our neighbours and our great wish is to make the colony air-minded'.[55]

The lives and achievements of these dozen or so pilots were different. The circumstances in which they traversed the Empire in the period 1928 to 1939 were diverse. Their motives, personalities and challenges varied. It is tempting, nevertheless, to compress their experiences, repress their individuality, and render them and their flying as products and statements of Empire, as its unappointed agents and as influences on Empire. The coincidence of late Empire with their determination, indulgence and daring in the sky is certainly hard to ignore. Caricatures are hard to resist.

By the 1930s, years of dreaming, aeronautical progress and political negotiating had, in theory, turned much of the British Empire into a theatre for long-distance flying. The alignment of tens of thousands of miles of Empire air routes had been settled; there was a basic infrastructure for refuelling and repair at RAF stations and at rudimentary landing grounds which were hacked, levelled and mown for commercial civil aviation.

The compilation of navigation maps and weather charts lagged behind. The sky was the new frontier, its nature unmapped: unpredictable winds affected fuel consumption (and therefore flying range and fuel loads); air density at different altitudes affected landing and take-off speeds (and therefore aircraft loading). Indeed, private flying to, from and in the Empire was axiomatically adventurous. It was exploration all over again. Haze and smoke affected visibility adversely and could obscure the railway lines and rivers used as markers. On the ground, heat and humidity slowed all manual work, including the basic task of refuelling. Even prepared landing surfaces contained surprises: tall trees and hillocks might conceal approaches; earth might be stony, sandy or waterlogged; tree stumps and rocks lurked beneath long grass. Finding a serviceable surface on which to make an emergency landing could be extremely difficult. Once down, it could be difficult raising an alarm and getting help. Those were the lessons from the pioneering Empire flights in 1920. The same lessons of surprise and danger emerged on private flights thereafter.

Ignorance of the difficulties could not explain persistent flying. Denial and recklessness might have explained some venturing. Financial greed, psychological need and boredom were considerations. Francis Chichester, for one, was simply not suited to a sedentary life. A sense of mission was either claimed by some pilots (or other commentators) or attributed to flights in retrospect. Thus, when Bentley flew from London to Cape Town in 1927, the London editor of the sponsoring Johannesburg *Star* newspaper explained that the trip would help to develop and popularise aviation among the people living along the vast spaces between the Zambezi and the Cape where rapid transport was essential to prosperity.[56] How Bentley regarded his flight was not mentioned.

Little, if any, of men's solo flying between the extremities of Empire in the 1920s and 1930s was concerned with just passing time and enjoying a novel experience. Sterner objectives led: reaching a destination faster than ever before was the most common. Women pilots, on the other hand, had a different way of 'being first' in the late Empire. Men had been the first people (and the only sex) to fly partnered or solo between many parts of the Empire, and could only trump one another on another 'plane'. At the start of the 1930s, however, women still had the opportunity to be the first of their sex to reach Empire destinations by air, or to follow an unusual route or direction. Once unique accolades had been awarded, of course, women followed men in a race to be the quickest – this time, the quickest person of any sex or nationality.

In 1928, two Irish women flew separately the 6,150 miles separating the British imperial capital and one of its extremities, Cape Town. Lady Mary Bailey and Lady Mary Heath travelled in different ways; the former was careless about her appearance, the latter stereotypically feminine. Along with a six-bore shotgun and fifty rounds of ammunition, she carried a tennis racquet, six pairs of silk stockings, six day-dresses, a fur coat and a tiny silver-fitted dressing-case. Flying across Africa was so safe, she said, that a woman could travel wearing a Parisian dress and keep her nose powdered the whole way. The cosmetics took precedence over shoes, boots, book and tennis racquet that Heath jettisoned in order to lighten her aeroplane while trying to cross high ground in East Africa.[57] Thoughtfully, she aimed them as close as she could to the scattered houses of settlers where she thought they might do someone some good. Like the caption writer for a published map of her African route, Heath supposed she had left civilisation behind. News reached even remote farms, however, and the press had not yet corrupted honest rural values. Reading about the abandoned tennis racquet in a newspaper, a farmer mailed it to 'Lady Heath, The Aerodrome, London'.[58]

The two Marys had independent means. Lady Heath was thirty-one when she flew across Africa. She had been married only a month to her second husband, Sir James Heath (seventy-five), a wealthy colliery owner, when they sailed to Cape Town with her new light aircraft in the ship's hold. For a second time, she had married for money: she was not born into wealth, and was brought up by her aunts after her father murdered her mother. She had her father's genes: fearless, wild and difficult to control. But she was bright, finished a science degree, was a superb athlete and served as a military dispatch rider. All her short life she was torn between craving (boyish) adventure and longing for (feminine) acceptance in the higher echelons of society. Revelling in her recently acquired title, the public adulation Lady Heath received in South Africa as the first woman pilot visitor massaged both needs. After a whirl of flying, speaking and social engagements, her trans-Africa flight took shape when she volunteered air support for an overland Cape-to-Cairo motor expedition. At the end of a three-month journey full of incidents, Lady Heath became the first woman to fly solo from Cape Town to London, and the first to fly solo from any British colony to London.[59]

Lady Bailey, a pupil of Lady Heath's, was the second wife of the rich and well-connected South African mining magnate, Sir Abe Bailey. She flew herself on a return trip between London and Cape Town in 1928. Aged thirty-seven, she was seeking fulfilment away from her role of society hostess, and escape from the prams into which she had put one new child for five successive years. The purpose of her outbound and return journeys could be masked by the socially acceptable wish to visit her husband (then in South Africa) and return to her children in London. She was in no hurry and spun out approximately 245 flying hours over 18,000 miles into a ten-month odyssey.[60] The stretch included a month in Leopoldville 'wrestling with the discomfort of poisoned mosquito bites' on her way back to England through West Africa.[61]

Alone at the controls of their respective open-cockpit aircraft, ladies Bailey and Heath had assistance from British men in the territories over which they flew, and also played the part of celebrity-women at hastily organised dinners and balls. Hospitality and aeronautical expertise on land were vital. At a time when both French and British authorities forbade women to fly alone across the Sahara, both Bailey and Heath were escorted by a male pilot (Flt/Lt Bentley) for that sector of their journeys. In regard to natural hazards, the protective instincts toward women were a residue of them being considered a weaker sex. Yet men too might die 'a slow agonising death by insect bites' in Sudan's swampy Sudd. Other dangers might indeed affect men and women differently. Whereas both men and women might be shot at by locals who did not distinguish between civilian and military aircraft, it was only women

Figure 2 Lady Bailey and British officers attending to her Moth in the Sudan (probably Khartoum), March / April 1928.

who were deemed tradable commodities with a value equivalent to two goats.[62] Only men were given 'goolie' chits, a refinement of the RAF document that promised a reward intended to dissuade captors from castrating downed airmen. While in Tripoli on his England–Australia solo flight in 1929, Francis Chichester took the precaution of getting an old acquaintance of his to write in Arabic that they had been boyhood friends and that he should be helped if necessary.[63]

As it happened, he did not need physical protection. But, like all those who flew imperially, Chichester did draw on British resources and networks overseas. The author of the Preface to his book was wrong in claiming that 'within the Empire no special gifts but energy and determination may win the prize of swift adventurous success'. On the contrary, Chichester himself acknowledged the help he received from British pro-consuls at Tripoli and at Benghazi (where he had an old school-friend), from the *Imperial Airways* mechanics and staff at Rutbah Wells (Iraq), Bushire and Jask (Persia), from the British aerodrome manager at Baghdad, the de Havilland mechanic at Karachi, and the British manager of an aerial survey company in Calcutta.[64]

Bailey and Heath were welcomed admiringly in London. The day after Lady Bailey's return in January 1929, when 200 people gathered to welcome her at Croydon, she was lunched at the smart Savoy Hotel. Lord Thomson (Secretary of State for Air), Cobham, de Havilland and Hinkler were present. The Royal Aeronautical Society, the Royal Aero Club and the Air League of the British Empire had invited about 100 guests. Lady Bailey gave a brief, understated talk. Later, she reflected more earnestly on the lessons of her flight, adamant that she had not started out with any imperial preconceptions or purpose.

Bailey admired the excellently run three-year-old Belgian and eight-year-old French commercial air services across West Africa. Whereas both operated British aircraft, it saddened Bailey that there was no British commercial air service in any of the British territories. It was 'disheartening', she wrote, that eight years since van Ryneveld's flight, Britain was not among the nations penetrating Africa peacefully by air. 'It may be merely the impatience of a woman', she added, but was it not time to stop quibbling about the cost of an All-Red route and consider instead 'the broader imperial aspect'? Her appeal for England to organise an air route from Cape Town to Cairo would have featured in her many articles and talks. One of her presentations, in March 1929, was to the Ladies Imperial Club. Brig. P. R. C. Groves (representing the Air League) and William Forbes-Sempill spoke at the same meeting.[65]

Lady Heath, also fêted in London, reached similar conclusions. She felt it a 'criminal shame' that gallant flying efforts had been wasted. Her May 1928 luncheon talk in maternal idiom (she referred to colonies as Britain's 'younger sisters') contained conventional ideas about imperial duty and useful overseas territory. We in England, she wrote later, do not realise that most of Africa is ours and that its great storehouse of mineral and agricultural wealth 'is ours if we like to take it and use it'. The intended crescendo of her talk was a reference to 'the two great assets of the British Empire – colonial settlement and British aviation'.[66] Colonial accounting was only a nuisance: having spent over £100 on telegrams, she was indignant at the Nairobi Colonial Secretary's bill for £20. The Colonial Office squashed the matter before the press and public could make merry over pettiness.[67]

The Bailey and Heath flights chipped away at male domination of long-distance flying. Beryl Markham edged the process onward in a flamboyant expression of female liberation.[68] In Kenya (and in England) her indiscreet, overlapping sexual flings included trysts with Jim Mollison and with the royal brothers the Duke of Gloucester and the Prince of Wales. The premises of the Royal Aero Club in Piccadilly (a bastion of privileged masculinity) proved handy for assignations. Her behaviour bewildered and scandalised the colonial memsahibs and prurient London society whose moral compass was jammed on female passivity and fidelity.

In 1932, aged thirty, Markham flew solo from Nairobi to London in seven days. Back in Kenya she worked freelance for an air charter company, performed for the local flying circus, and flew as an airborne elephant scout for hunting parties whose members included the Americans A. Vanderbilt and Ernest Hemingway.[69] She also worked as a pilot for a visiting British parliamentarian and the son of Capt. F. E. Guest (Secretary of State for Air in the early 1920s), himself a Kenyan landowner. In 1935/36, while preparing for her greatest aviation feat, Markham replaced Amy Johnson (see below) as pilot for the fastidious clientele of a French air cruise firm. Four years later she became the first woman to fly across the Atlantic from east to west. Adored in America, she returned to the British world where misinformation, jealousy and outrage restrained public acclaim. The last of Britain's notable solo women flyers slid into obscurity. Organised civil aviation, less high-spirited, had become the standard bearer of air Empire.

Adventuress and empress

Sandwiched between the Bailey/Heath and Markham flights was one which pilot, public and press imbued with even more explicit and penetrating imperialism. On 5 May 1930 the daughter of a notable Hull businessman (a herring importer) set out on a solo flight to Australia. A pupil at a London flying club, she had 100 hours' solo flying experience. She was one of sixty women pilots among 2,000 in Britain. Her second-hand Moth (*Jason*) was part-funded by Lord Wakefield (the philanthropic oil magnate) on Sir Sefton Brancker's recommendation.[70]

Neither the public nor anyone in aviation offices and Fleet Street paid much attention. But after Amy Johnson (twenty-seven) reached Darwin in northern Australia nineteen days later (coincidentally on Empire Day), the clamour was astonishing. In London, Francis Yeats-Brown rushed into print, telling readers of the weekly *Spectator* that flying was a sign of British vitality: 'old gentlemen in London clubs might cease to drivel of our decadence if they saw how young England yearns for the air, where we must renew the greatness we won through centuries of seafaring'. He ended by acknowledging Lord Wakefield's readiness to take risks and back flying adventures, and by praising the *Daily Mail* for helping consistently for more than twenty-five years.[71]

Johnson had not flown fastest, but she was the first woman to get herself 'down under'. Celerity was not a prerequisite for celebrity. Hers had been an exhausting, touch-and-go adventure, propped up by luck and kindness along the way. At Rangoon, for example, wing fabric was repaired with shirts that 'wives of Englishmen' had sewn from war-surplus aircraft fabric. Only spoilsports would have paid more attention to chance than to skill and fortitude. Among 200 telegrams were three from Britain's King, Queen and Prime Minister. Delirious Australians turned out in their thousands to welcome her. The crowd at Darwin was considered the largest ever assembled there. Her arrival at one small town was said to have been the first-ever event of interest there. Approximately 75,000 people greeted her at the airport in Sydney.

Australians coined more than a dozen titles for Johnson, one of which was 'Ambassadress of Empire'. The phrase echoed the view of a British woman M.P. that Johnson had demonstrated persistence of 'qualities which in the past made the British race'. Gifts accompanied the shower of adulation during Johnson's six-week tour of Australia. As a quirky token of British supremacy she was presented with a scrap from the aeroplane flown in the First World War by the German air ace, Baron von Richthofen. During a theatre show interval her plane was wheeled onto the stage accompanied by 'a bevy' of women pilots.[72] Hurrahs and cheers resounded everywhere, not least in the chorus of Australia's 'National Official Welcome Song'. Its two main verses were about the gallantry of the Yorkshire lass who was true to 'high traditions' and who personified 'British pluck'. Johnson was 'British born and British bred'. As if an inevitable consequence, she was also 'a battler through and through'. Grander still, her flight meant 'linking nations as she sped a sailing through the blue'. Her attendance at a conference of the Women's Christian Temperance Union in Brisbane sealed the angelic image.[73] A popular song dubbed Johnson 'Queen of the Empire'.

Albeit inaccurate, it made good copy to write about Johnson being the first daughter 'sent forth' to do great deeds for Britain. Her flight was certainly transgressive, and it was bound to be disruptive at the destination. It did not permanently transform the gendering of Australian society, nor did it reform Australian suspicions about their imperial subservience. Individual and contradictory interpretations and inscriptions created a maze of plural and diverse popular responses. Johnson herself was firm only about the political links between flight and Empire. At Brisbane she declared that she stood for Empire, and that aviation and Empire stood together. At Melbourne airport she told 15,000 spectators to 'dream dreams and see great visions'. These, she implied, would 'breed a race of airmen comparable to Drake's seadogs' whose skyways would bind the Empire closer. Johnson herself must have read or heard often the mantra she regurgitated: 'Our great sailors won the freedom of the seas, it's up to us to win the freedom of the skies'.[74] Breed? Race? Airmen? Our? Freedom? The oratory was pure 1920s propaganda.[75]

If they disagreed about what her flight meant about modernity, Australians agreed that Johnson's journey drew Australia and Britain closer together geographically. Her flight also reinforced imperial commonalities and boundaries. Johnson spoke of feeling 'at home', and wondered if she may wake from a dream to find herself instead 'in some awful place with blacks all around me'. A shared imperial identity did not require that the centre and extremity of Empire dissolve, but that they were differentiated from non-imperial places. Indeed, these were erased from some cartographic representations of her trip. The Orient was reduced to an undistinguished, menacing space. In keeping with the sentiment, annotations on the route map that was published in the weekly London glossy paper, the *Sphere*, noted that wild men in Timor had rushed out of the bush, yelling and brandishing weapons. Standing firm in front of strangers whom Johnson herself labelled 'grinning savages' won her the respect of the Returned Soldiers' League in Australia, which regarded her as embodying spirited British wartime 'womanhood'.[76]

Johnson arrived back at Croydon in August 1930 after travelling by sea from Australia to Cairo. She had been refused passage back from Karachi on the ill-fated *R101* airship (an all-male complement was required),[77] but the Chairman of the P&O shipping line gave her complimentary first-class passage and free carriage for *Jason*. The final lap onward to England was with *Imperial Airways*. On a public holiday, 20,000 people turned out at Croydon.[78] The Air Ministry had learned lessons from the scrimmage when Charles Lindbergh landed there: British self-restraint had vanished, much to the amazement of a German journalist who reported bedlam surpassing the Wembley Cup Final and the annual university boat race.[79]

The stage would be managed better for Johnson. Enclosures were set aside for the public, invited guests were given tickets, chairs were hired. The Automobile Association assisted with traffic control, additional space was made for car parking. An RAF boxing ring doubled as a raised reception stand. Lord Thomson, Brancker and Wakefield were among the dignitaries who perched on a dais awaiting the *Imperial* flight, which was delayed by three hours. George Woods Humphery, *Imperial*'s Managing Director, pressed an airline steward into serving them refreshments. Eventually the airliner and several private light aircraft escorts came into view. It was a unique homecoming. Thomson's welcome and Johnson's reply were broadcast by BBC radio. Spectators stared and cheered when Johnson passed them on her car tour around Croydon's perimeter.[80]

The frenzied reception spilled out further. Appreciative crowds lined the road all twelve miles and two-and-a-half hours drive from the airport to the Grosvenor House Hotel in West London. There, Johnson's scrambled public utterances referred to more than mere machinery: 'I want to show by my flying how much I owe England and its people, how glad I am to be back home, how proud I am to be a member of our own great Empire, and how deep is my gratitude to you all.' The socialite editors of the *Sphere* were less captivated by her patriotism than by her appearance and ensemble. With staggering insensitivity to the significance of the moment, they focused on 'her clear cut features, her sculptured hair, her coat with its military collar . . . her emerald green handbag, hat in hand'.[81] A photograph of the heroine filled the cover of one edition of the magazine. Having once made do with sales, typing and secretarial jobs, the Sheffield university graduate was suddenly a public figure. Her gifts included a car from Sir William Morris, a speedboat and a new Moth.

In the predominantly male world of aviation, Johnson's gender mattered, or was made to. Her class was also of some account. Nobody fussed about her unsophisticated Humberside accent. On the contrary, many people warmed to her humble origins and her girl-next-door image. Unlike the Duchess of Bedford and Lady Bailey, Johnson was not connected with High Society. Yet she supplanted film stars and debutantes. 'No English-woman outside the Royal Family had ever known such intense and prolonged adulation', wrote one historian. William Courtenay, who adopted the roles of Johnson's guardian and counsellor, noted that women felt she had struck a blow for them. He shielded her from lesbians lusting after a lioness clad in leather flying gear, but allowed her to jolt the male establishment.[82] When welcoming Johnson, Lord Thomson might well have reflected that if a 'mere girl' could fly alone to India and beyond in record time despite little experience, then it had to be possible for a professional crew to fly an airship to India and back regularly, especially with all the resources of the State behind them. The First Lord of the Admiralty remarked at the annual Empire Day dinner in London in 1930 that it was impossible to consider the matter of shortening distances and communication within the Empire without thinking of Johnson's 'intrepidity and courage'.[83]

The establishment responded to Johnson's achievement by making her a Commander of the Order of the British Empire, reciprocating her pride in it.[84] She was elected an honorary member of the year-old Guild of Air Pilots and Air Navigators of the British Empire. Johnson also joined the ranks of the illustrious in being invited into the BBC. In a radio broadcast she expressed the hope that the youth of Britain would join her in furthering the ideal of making their country 'a glorious country in the air'. Great Britain, she continued, 'is ready to make a decisive bid for world supremacy in the air. Please all help, by talking, by example, and by flying.' In Hyde Park, a crowd of 100,000 saw Johnson accept the *Daily Mail* award of £10,000 and a gold cup to commemorate her 'British courage and endurance'. After a triumphant drive through London she arrived at the Savoy Hotel for a lunch given in her honour by the newspaper. Two hundred and fifty meals were served to a predominantly youthful audience. Johnson sat beneath a floral model of *Jason*. Selfridges' Oxford Street department store, which had an aviation section selling air route maps, flying gloves and goggles, was quick to market a miniature ladies brooch of the plane.[85] It may have been worn as a badge of gender, wealth or hobby.

Beyond these environs, Johnson's flight 'lit a million imaginative lamps in the hearts of a million men and women'. Most of these were British, but even in far-off India a small group of women had begged her just to touch their hands. She did so cheerfully, a gesture that *The Times* considered immensely generous and gracious in view of her tiredness. Even before she returned to England, the Women's Automobile and Sports Association was among the first to donate funds to buy Johnson a new plane. Among British people spurred to join flying clubs as learners were women, who began diluting men's domination of private flying. Despite male condescension, Britain's first all-women's 'flying meeting' was held in September 1931 at Sywell aerodrome, Northampton. Another was held in August 1933 at Maidstone. London women's interest in flying was sustained also by the aviation section of the women-only Forum Club, out of which sprang the Women's Aero Club. *Imperial Airways* exhibited at the Club in April 1934. If there were sexual politics afoot, Johnson did not align herself to an exclusionary wing: in 1932 she became the first woman ever invited to the Royal Aeronautical Society's annual council dinner.[86]

Johnson's efforts were celebrated in a torrent of articles and in popular songs. Those with imperial resonance included one composed by her musician uncle. It ended with the words 'Salute the Queen of the Air, in the British Empire's name'. Courtenay's own 1933 poem, 'Ode to Amy', encapsulated the haughtiness of Empire and the appropriation of Empire aviation enterprise by the English. The poem welcomed her home to an England, which, unlike places abroad (he wrote), was 'safe from decay, corruption or neglect'. He registered her flight as a sign of fighting another day for England, and as the epitome of courage and steadfastness – England's 'ancient virtues'.[87]

The air exploits of the Hull heroine did not end with her Australia trip, but no subsequent flight roused quite as much interest. Late in 1932 she broke the record for flying between England and South Africa in both directions. A Sunday newspaper told its English readers that she had brought Cape Town as near to London as Edinburgh was for their grandfathers.[88] Her flight, she said on a BBC broadcast from Cape Town, had been partly to maintain British prestige in the air. She smashed the records set the previous year by Lt-Cmdr G. Kidston and Lt O. Cathcart-

Jones (six days, eleven hours), by Gordon Store and Peggy Salaman (five days, six hours) and by her husband Jim Mollison (four days, seventeen hours, in 1932). Johnson tackled the West Africa coastal route to and from Cape Town (700 miles shorter than the easterly, continental airway) in 1936.

Her African adventures did no harm to either her popularity or to the promotion of Empire aviation. When Johnson toured British provincial cities her aeroplane was exhibited for a week at Lewis's department stores in places like Leicester, Liverpool and Manchester. Lewis's also laid on a lunch in her honour, to which distinguished citizens were invited. In the cheering crowds, women wept and schoolchildren gawked in awe and adoration. Girls who read the *School-Days* weekly magazine and had retained the free stand-up photograph of Johnson in an August 1930 edition saw their heroine in the flesh.[89] Schoolgirl visitors to Madame Tussaud's wax museum in London voted her one of the three women they most wanted to emulate, second to nurse Edith Cavell and ahead of Joan of Arc.[90]

The Johnson fairy-tale has always been a little tarnished. Even in the 1930s Johnson was accused of avarice and of being complicit in a sensational but false press report about her terror in a dark forest among ogres when she crash landed near Warsaw in 1931.[91] As 'adventuress', her impositions in the Netherlands East Indies created imperial embarrassment.[92] Recent scholarship has continued to probe the hagiography and mythology, investigating the cult and impact of the aviation heroine.[93] Johnson manipulated her image, straddling masculinity and femininity. Her marriage (albeit brief and unconventionally modern) facilitated cultural acceptance. It has been observed how she (and her husband) operated as self-promoting celebrities in a carefully orchestrated media campaign, and how she legitimised her exploits by linking them to approved imperial tasks.[94] In doing so she piloted her way through contradictions, for her flight was also used to reinforce divisions between imperial territories and peoples. Yet the dominant conception of coherent Empire was unruffled; prevailing notions of femininity were shaken only briefly. In examining the social re-production of the 'aviatrix', a query has arisen about the liberation that Johnson won for herself and other women. Out of the air, on land, it has been suggested that Johnson's 'freedom' was notional, constrained by institutions and discourses that favoured men.[95]

Britannia's other daughters

Johnson was the first, but neither the only nor the last woman to soar solo between the extremities of the British Empire. Late in 1934, twenty-seven-year-old Freda Thompson became the first Australian woman to fly solo from England to Australia. She was successful despite damaging her aircraft in Greece.[96] Lores Bonney flew in the reverse direction in 1933 on a solo 'flight' that included one stretch of sea travel. Two years later a New Zealander, Jean Batten (twenty-six), became the first woman to fly solo in both directions between Australia and England. She masculinised her unmarried status as a choice 'between matrimony and career'. Her verdict: 'ambition claimed me'.[97]

Batten bettered Amy Johnson's exploits on the England–Australia route in the 1930s.[98] Fired by news of record-breaking long-distance flights, and by meeting Kingsford Smith, the fiercely determined Jean and her mother sailed from their New Zealand home to England in 1929 specifically to take flying lessons. After two abortive starts, and with financial help from Wakefield among others, she finally became the second female Empire citizen to fly solo to Australia. She landed her Moth at Darwin in May 1934.

Too competitive to have postponed her arrival by eleven hours to coincide with Empire Day, Batten nevertheless made a great deal of the imperial significance of her flight. In Sydney, she told a welcoming crowd that she hoped her flight would 'help to strengthen the great bonds between all the Dominions and Colonies of our great Empire'. It was a sweeping hope for a flight connecting only three imperial territories. And despite the start of several commercial Empire air services, and the imminence of others (few of which had relied on the heroics of solo pilots), Batten was convinced that she was contributing to the development of Empire air routes. In almost every one of 150 speeches she gave in New Zealand, she was even more emphatic that she had actually 'been able to strengthen the great bonds of friendship not only between England, our Motherland, Australia and New Zealand, but between all the Dominions and Colonies of the Empire'. Similar imperious sentiments would not have been far from the minds of the speakers who eulogised her tediously with excruciating and extravagant sycophancy. The Mayor of Auckland was some way back in the queue when remarking merely that she had 'brought honour to the British Flag, added lustre to womanhood, and enhanced the traditions of the British race'.

Batten's flight back to England in seventeen days in May 1935 (to watch the King's jubilee pageant) meant that she became the first woman to fly in both directions between motherland and homeland. The celebrity was fêted by the aeronautical establishment. A flight across the South Atlantic to Rio de Janeiro and Buenos Aires later in the year enhanced her reputation. Being made a Commander of the Order of the British Empire in 1936 topped the acclamation; the Air Ministry bridled at the audacity of the New Zealand Premier's request that she be made a Dame of the British Empire. The decidedly unofficial title 'Empress of the Air' became Batten's after her eleven-day flight from England to New Zealand later that year. By then, however, the twenty-seven-year-old's imperial services were more obviously secondary to her search for personal glory. Nevertheless, it has been possible for her biographer to write that her arrival in Auckland surpassed even a royal visitation and 'will remain for ever one of the greatest events in New Zealand history'.[99] Personal, national and imperial elevation went hand in hand.

As in Johnson's case, the clamour surrounding Batten was astonishing. Both women's heroic reception contrasted with the chill which Lorres Bonney felt after her epic flight. In her hometown of Brisbane, women's organisations refused to comment on grounds that they were involved in business not social matters. Bonney's biographer reflects that 'most Queensland women, tied to husband, home and children, resented her freedom and the notoriety it had brought her. In an era when women's place was firmly established in the home, most would have believed it to be Mrs Bonney's also'.[100] Townspeople who frowned on the deceptions that enabled Mrs Bonney to take up flying without her husband's knowledge were meaning to address her rebellious behaviour, but in so doing they highlighted the exclusionary culture of flying.

Notes

1 W. H. E. Thomas (ed.), *The Flying Clubs' and Schools' Year Book* (London, 1934).
2 Multiplying 1930s prices by fifty is a handy way of approximating equivalent prices in 2010.
3 R. McKibbin, *Classes and Cultures: England 1918–1951* (Oxford, 1998).
4 S. McKay, *de Havilland DH.60 Moth* (Hinkley, 2005).
5 McKay, *de Havilland.*
6 *Aeroplane* (30 March 1932), pp. 552–60; A. Goudie, *Wheels Across the Desert* (London, 2008), p. 143.
7 F. Chichester, *Solo to Sydney* (London, 1982).
8 *Flight* (31 October 1934), p. 1094; (28 March 1935), p. 336.
9 J. Williams, 'The upper class and aeroplane sport between the wars', *Sport in History*, 28 (2008), 450–71.
10 *Flight* (11 October 1934), pp. 1062–3. For more detail, and images, see D. Cluett, *Croydon Airport: the Australian Connection* (Sutton, 1988).
11 *Cape Times* (3, 4 March 1926); McKay, *de Havilland.*
12 On the Hoare flight see Pirie, *Air Empire*, pp. 102–7.
13 *Journal of the Royal Aeronautical Society*, 31 (1927), 910–11; T. Boughton, *The Story of the British Light Aeroplane* (London, 1963).
14 *The Times* (28 January 1933), p. 7. Richard's motor cars and aeroplanes are the core of the present-day Shuttleworth Collection at Old Warden Aerodrome.
15 *Bystander* (1 May 1934), pp. 234, 236.
16 K. C. G. Dower, *Amateur Adventure* (London, 1934).

17 *Manchester Guardian* (29 December 1934), p. 13; *Flight* (31 January 1935), p. 132.
18 A. Henshaw, *The Flight of the Mew Gull* (London, 1980), p. 38.
19 *Cape Times* (3 October 1931).
20 *East Africa* (23 April 1931), p. 995.
21 A. J. Cobham, 'My flight round Africa', *Elder Dempster Magazine* (1928), 84–90; A. J. Cobham, *Twenty Thousand Miles in a Flying-boat: My Flight Round Africa* (London, 1930). The distances of Empire flights were reported inconsistently. Faulty measurement, approximation, ignorance, sloppy checking and aggrandisement led to rounding and exaggeration.
22 J. Carberry, *Child of Happy Valley* (London, 1999).
23 *East Africa* (28 June 1928), p. 157; *Daily Mail* (28 June 1928).
24 *The Times* (12 April 1930; 1 July 1931; 2 October 1935; 21 December 1936); *Flight* (10 July 1931); *Aeroplane* (7 September 1938).
25 *The Times* (24 May 1933).
26 *Listener* (22 September 1938); *Daily Mirror* (15 July 1933), p. 5.
27 *Airways* (February 1928); I. Mackersey, *Jean Batten: the Garbo of the Skies* (London, 1991).
28 *Airways* (August 1929); *The Times* (10 June 1929; 4 October, 3 December 1930); *Flight* (11 April 1929), p. 302.
29 *The Times* (2 March, 6 April 1931).
30 P. B. Maling, *Wanderlust: the Life of Aline Barton, Pioneer New Zealand Airwoman* (Christchurch, 1988).
31 *Flight* (11 December 1931); (29 January 1932), p. 95; *The Times* (2 March, 6 April, 3, 6 November 1931; 18 January 1932).
32 *The Times* (4 February 1935); *Aeroplane* (13 November 1935); *Sphere* (October–December 1935).
33 Mackersey, *Jean*; *The Times* (7, 10 March 1932).
34 A. Sanger, *Margery Durant Goes to Africa, 1931–2* (privately published, 2009); C. LaJotte, 'LaJotte flies in Africa!', *Western Flying* (April 1936).
35 *The Times* (28 March 1931; 29 August, 12, 29 October 1932).
36 *Sydney Morning Herald* (6 October 1932), p. 9; *Mercury* (Hobart) (18 January 1933), p. 5; *Brisbane Courier* (21 March 1933), p. 12.
37 McKay, *de Havilland*, p. 96.
38 V. Bruce, *Nine Lives Plus* (London, 1977).
39 *The Times* (4 May 1931; 29 August, 12 October 1932).
40 *The Times* (3 January 1935).
41 Dower, *Amateur*, p. 23.
42 G. Whittell, *Spitfire Women* (London, 2009); *The Times* (18 January, 27 October, 11 November 1932; 18 January 1933), p. 9; *Friend* (Bloemfontein) (30 November 1932); *Star* (Johannesburg) (16 January 1933).
43 E. Beinhorn, *Flying Girl* (London, 1935).
44 Speculation that the Duchess committed suicide by flying out to sea (e.g. M. Russell, *The Flying Duchess* (London, 1968)) has been countered by the suggestion that she simply made a navigational error. L. Curtis, *Winged Odyssey* (Walton on Thames, 1993).
45 Russell, *Flying*; J. Gore, *Mary, Duchess of Bedford, 1865–1937* (London, 1938).
46 *The Times* (7 September 1928), pp. 12–13; Russell, *Flying*; *Sphere* (17 May 1930), p. 361; *Air* (June 1930); Duchess of Bedford, *To the Cape and Back in Twenty-and-a-Half Days* (Air Press Agency, 1930).
47 Gore, *Mary*.
48 *East Africa* (11 December 1930), p. 421; (8 January 1931), p. 550.
49 British Library (London) (BL), OIOC, L/PO/1/31 (ii); *Flight* (23 May 1930); *Air* (May 1930).
50 *Aeroplane* (26 March 1930), (14 May 1930), p. 902, (4 June 1930).
51 S. H. Smith, *Indian Airways* (Part 3) (Sutton Coldfield, 1930); Boughton, *Story*; *Airways* (July 1930), p. 385; Political Diaries of the Persian Gulf (PDPG) (December 1930; January 1931); McKay, *de Havilland*.

52 *Flight* (11 October 1934), pp. 1062–3.

53 D. Jones, *The Time Shrinkers: the Development of Civil Aviation between Britain and Africa* (London, 1971).

54 Markham's autobiographical account of her flying experiences in *West With the Night* (Boston, 1942) preserved imperialist masculinity and imperial individualism: S. Smith, *Moving Lives: Twentieth-Century Women's Travel Writing* (Minneapolis, 2001), pp. 109–10; 117–18.

55 Carberry, *Child of Happy Valley*, p. 15.

56 *The Times* (17 August 1927).

57 S. M. Heath, 'My Cape Town to London flight', *Asia* (February 1930), pp. 81–7; 'Sunstruck in the air', *Asia* (March 1930), pp. 200–207; 'On to Cairo', *Asia* (April 1930), pp. 284–90.

58 *Sphere* (19 May 1928); *The Times* (24 May 1928); *Air* (July 1928).

59 L. Naughton, *Lady Icarus: The Life of Irish Aviator Lady Mary Heath* (Dublin, 2004).

60 M. Russell, *The Blessings of a Good Thick Skirt: Women Travellers and their World* (London, 1996); J. Falloon, *Throttle Full Open: a Life of Lady Bailey, Irish Aviatrix* (Dublin, 1999); *Independent* (London) (24 March 2000), p. 6.

61 *Times* (29 January 1929), p. 15.

62 H. E. Brittain, *Wings of Speed* (London, 1934), pp. 162–3; *Sphere* (19 May 1928); *Air* (July 1928), p. 24.

63 Chichester, *Solo*, p. 88.

64 Chichester, *Solo*.

65 Falloon, *Throttle*, pp. 161–2, 171; *The Times* (8 March 1929).

66 *Air* (July 1928), p. 24; S. M. Heath and S. W. Murray, *Woman and Flying* (London, 1929), pp. 125, 157, 217.

67 Naughton, *Lady Icarus*, p. 156.

68 This and the next paragraph draw on E. Trzebinski, *The Lives of Beryl Markham* (London, 1993).

69 Light aircraft were not used as wildlife shooting platforms, but as tracking devices they did expedite game chases using motor vehicles. See J. M. MacKenzie, *The Empire of Nature: Hunting, Conservation and British Imperialism* (Manchester, 1988); P. C. Mitchell, 'Aeroplanes and African fauna', *Spectator* (1929), 502–3; L. Montagu, 'Aviation as affecting India', *Journal of the Royal Society of Arts*, 67 (1919), 543–53.

70 Although Johnson contemplated naming her aeroplane *Spirit of England*, the name which stuck was *Jason* (a contraction of the family fish shop name). E. Millward, *Women in British Imperial Airspace, 1922–1937* (Montreal, 2008), p. 83.

71 *Spectator* (7 June 1930), p. 933.

72 *The Times* (7 June 1930), p. 11; (17 June 1930), p. 13; J. Thomas, 'Amy Johnson's triumph, Australia 1930', *Australian Historical Studies*, 23 (1988), 72–84.

73 National Women's Library (London Guildhall University): E. N. MacCulloch, 'Australia's National Official Welcome Song to Amy Johnson' (London, 1930); *Argus* (Melbourne) (31 May 1930), p. 25.

74 M. Paris, *From the Wright Brothers to Top Gun: Aviation, Nationalism and Popular Cinema* (Manchester, 1995), p. 103; Thomas, 'Amy Johnson's triumph'; J. Lloyd, 'The impossible aviatrix', *Australian Feminist Studies*, 15 (2000), 137–52.

75 Compare Lord Thomson's clarion call in his *Air Facts and Problems* (London, 1927), pp. 144–67, 188.

76 *Sphere* (31 May 1930), pp. 436–8; *The Times* (17 June 1930).

77 J. Gardiner, *The Thirties: an Intimate History* (London, 2010), p. 695.

78 *Flight* (6 June 1930), p. 607; (8 August 1930), pp. 882–5.

79 M. Eksteins, *Rites of Spring* (London, 1989), pp. 246–7.

80 National Archives (Kew) (NA), AIR 2/371.

81 *Sphere* (9 August 1930).

82 W. Boase, *The Sky's the Limit: Women Pioneers in Aviation* (London, 1979); W. M. M. Courtenay, *Airman Friday* (London, 1937), p. 111.

83 P. G. Masefield, *To Ride the Storm: the Story of the Airship R101* (London, 1982), p. 265; *The Times* (27 May 1930), p. 11.
84 Despite the 'hardihood', 'fortitude' and 'gallantry' of Johnson's 'rashly venture', its proof of 'the enterprise of indomitable British womanhood' and the 'new hope for the Empire', it was easy to balk at a knighthood for feminine valour: 'Sir Amy' just didn't sound right. *Pacific Affairs* (September 1930), p. 826.
85 *Daily Express* (4, 7 August 1930); *Listener* (13 August 1930), p. 240; *Air* (August 1930); *Flight* (15 August 1930), pp. 916–19; D. Cluett, et al. (eds), *Croydon Airport: the Great Days, 1928–1939* (Sutton, 1980), p. 162; Boase, *The Sky's the Limit*; C. B. Smith, *Amy Johnson* (London, 1967).
86 Courtenay, *Airman*, p. 16; V. Moolman, *Women Aloft* (Alexandria, VA: Time-Life, 1981), p. 103; *The Times* (28 May 1930); *Journal of the Royal Aeronautical Society* (December 1932); *Air and Airways* (March 1933), p. 402; *Aeroplane* (11 April 1934; 14 December 1938).
87 Courtenay, *Airman*, p. 102; Smith, *Amy Johnson*, p. 229.
88 *Observer* (20 November 1932).
89 *The Times* (19 November 1932), p. 12; Courtenay, *Airman*; *Woman's Weekly* (9 August 1930).
90 Gardiner, *The Thirties*, p. 694.
91 D. Luff, *Mollison: the Flying Scotsman* (Washington D.C., 1993), pp. 162, 205; M. Gillies, *Amy Johnson: Queen of the Air* (London, 2003), pp. 230–1.
92 Millward, *Women*, pp. 101–8.
93 See, for instance, R. Blythe, *The Age of Illusion* (Harmondsworth, 1963), pp. 83–102; Gillies, *Amy*.
94 B. Rieger, '"Fast couples": technology, gender and modernity in Britain and Germany during the nineteen-thirties', *Historical Research*, 76 (2003), 364–88.
95 Thomas, 'Amy Johnson's triumph'; Lloyd, 'The impossible aviatrix'.
96 *The Times* (1 October, 7 November 1934).
97 J. Batten, *My Life* (London, 1938), pp. 92–3; Millward, *Women*, pp. 151–83.
98 Millward, *Women*, pp. 117–50.
99 Mackersey, *Jean*, pp. 162–3, 170, 171, 247; Millward, *Women*.
100 *The Times* (5 October 1932); (1 October, 7 November 1934); T. Gwynn-Jones, *Pioneer Aviator: the Remarkable Life of Lores Bonney* (St Lucia, 1988), p. 116.

3

Seeking supremacy

By its very nature, flying was relatively high and relatively fast. At first, pilots set height and speed records unintentionally, but later they had to pursue them deliberately. Two such record-setting flying events in the British Empire caught the public imagination in the 1930s. Both presumed imperialism. Both projected imperialism. Their geographies were significant: they occurred at the limits and edges of Empire. Both were planned, not spontaneous. The competitive element was organised and transparent. Pilots competed against the elements and against one another. The events were not leisured. Entry was not a free-for-all. Routes were restricted. One event was the first flight over Mount Everest. Its promoters were right-wing and ardent imperialists enamoured by science and aeronautical technology. For them, this was 'a glorious story in a troubled time' when Britain's political leadership appeared supine and inclined to betray and lose the Indian Empire.[1] The second event was an air race from Britain to Australia.[2] Both events were exclusively male. Both occurred at a time when organised Empire flying was already established and when individual record breaking was largely over. Empire was the geographical backcloth and safety net: both events were flag waving but the pilots muted the imperial motives.

Being fastest

In February 1928 Bert Hinkler flew a light aeroplane single-handedly from Britain to Australia (he called it 'the promised land') in fifteen days. The Australian clipped twelve days off the Smiths' 1920 time. He consigned Parer and McIntosh's infamous 'concurrent' seven-month flight in 1928 even further into oblivion.[3] The ex-RAF squadron leader made his way on a 'mount' with the aid of *The Times* topographic atlas. Hinkler would have known how daylight flying hours shortened as he flew east, but his plans were upset by the Egyptians who barred him for failing to give fifteen days' advance notice of his flight, and by a Palestinian medical officer who insisted he present himself for an inspection parade. There had been rather greater obstacles beforehand: in England it had been almost impossible to raise funds for the flight. Lacking Cobham's charisma and persuasion, Hinkler thought it would have been easier selling rotten fruit. Advance corporate sponsorship was risky; retrospective product advertisement was safe. Fine sentiments about imperial interests that might be served by individual endeavour did not always write cheques.

Hinkler was not the first to fly to Australia, but at 100 m.p.h. he was the fastest. His solo flight was a first in its own right, and was commended for its imperial significance and personal courage. His effort had additional valency. He perpetuated interest in the possibilities of affordable air links between Britain and Australia. He also set a new standard in the public mind: by comparison, the proposed England–Australia timetable of fifteen days on an organised, commercial service flown by a multi-engined airliner appeared sluggish.[4]

This was public relations *Imperial Airways* did not need. Not only were commercial air links exposed to criticism, they were being overshadowed by a populist reinterpretation of their significance. Cobham insisted that rapturous welcomes celebrated the link with Britain, not pilot effort, but in Hinkler's case many clapped more for man and country than for Empire. *Punch* magazine made the point in a cartoon captioned 'Australia Advances'. A dancing kangaroo mascot waved an Australian flag and sang 'Hinkle, Hinkle, little star! Sixteen days and here you are!' The flier was welcomed by a crowd of 1,000 at a reception in Sydney town hall. The Australian public took the modest, no-frills airman to their heart. The son of a working man, he was the kind of hero with whom they could identify. Men everywhere sported 'Hinkler Homburgs' (his favourite headgear) and women the equivalent 'Aviator' hat. Couples danced the 'Hinkler Quickstep', and his name was incorporated in cookbook recipes; even a new strain of flower was christened the 'Hinkler Dahlia'.[5] The Empire was the wallflower at the party.

Imperialism was not prominent either (except in as much as Africa was said to be 'all wild beasts and savages') when, in 1931, Gordon Store and the allegedly nineteen-year-old Peggy Salaman broke the standing record of six days' flying time from England to Cape Town.[6] Store, then twenty-five, an assistant flying instructor at the London Aeroplane Club, would have welcomed the fee he was paid, but also saw the trip as an opportunity to visit his South African birthplace, Kimberley. Trainer and pupil triumphed in an aeroplane named *Good Hope*, a birthday gift from Peggy's mother. She elaborated that Peggy went to a Christian Science school and that both of them had 'complete faith in divine providence'. Back in London, at a dinner arranged by the Air League of the British Empire, Salaman told 200 guests that her job had been to relieve Store from the straightforward flying, and to look after customs formalities, the log book and telegrams. The two trophy lion cubs she bought at Juba accompanied her on part of the onward flight to Cape Town and then on her sea voyage back home to England.[7]

When, in 1931, Lt Cmdr Glen Kidston (thirty-two) set out to be fastest between England and the Cape, his flight did have some imperial purpose: it was to embarrass *Imperial Airways*. His six-and-a-half-day flight from England to the Cape (with a crew of two in a single-engined, American-built monoplane) lowered the record by two days and beat *Imperial*'s projected outbound schedule of eleven days. Indications were that the flying time could be reduced to nine days once pilots became familiar with the route, if night flying was introduced, and if mails and passengers were separated.[8]

Born into a very rich family, Kidston retired from the Royal Navy active list aged twenty-nine and had more time to race motor-cycles, cars and speed-boats. It is an open question whether he just learned to stress aeronautical utility and minimise romance, or whether he was sincere. He could certainly say the right things. Air transport, he once remarked, would 'play a very important part in the blissful state of . . . modern civilisation'. He switched from boyish delight in machines and speed to interest in the developmental potential of aviation partly because of his experience flying to East and Central Africa for gun and camera safaris. In 1928 he had been one of a party of three (excluding pilot Drew and a mechanic) photographed leaving Croydon on a hunting trip to Africa. His death in a Moth air crash within a month of setting his 1931 record ended that dream.[9]

Air disasters – and obituaries – often brought imperial flying into focus. After an air speed-record bid starting on the Solent in 1928 ended in tragedy (killing a South African pilot who had been responsible for some ground arrangements during the RAF Cairo–Cape flight that year), the shapers of public opinion tried to ensure that the accident did not dim public spirit more than momentarily. In a eulogy about his legacy and honourable death, the leader writers of *The Times* reminded readers that the RAF exercise was 'no mere sporting adventure': a serious experiment in the improvement of air navigation and air transport, it was made in the interest of the Empire with official permission and encouragement.[10] The *Spectator* mourned the loss of pioneer aviators, but did not counsel safety first. Instead, readers were told that Britons owed the glory they had as a race to pioneers who dared the impossible. The spirit of the Elizabethans was never more alive than in contemporary England, the paper added; 'there is no conquest with such a direct significance to the inhabitants of these islands than that of the air'.[11]

Private flying in the Empire became less and less about imperial achievements as the years passed. The Britishness of the pilots and their machines remained, but tearing between far-flung points of the Empire (especially in the same longitude so nightfall did not artificially shorten the 'flying day') was about using known routes if not always planned landings. In October 1935, Mrs Jill Wyndham and her flying instructor, David Llewellyn, flew from London to Cape Town hoping to beat the record, but were frustrated by a forced landing in a rice field (where the lamps of an Arab funeral procession guided them in). They did set a new record for the return leg to London. Llewellyn then returned solo to Johannesburg in March 1936.[12]

Two years later, Betty Kirby-Green (a licensed pilot, thirty-one) and flying officer Arthur E. Clouston (thirty) set a flying record for the 14,700 mile round trip to the Cape, landing at Croydon within six days of having left.[13] Clouston was no upstart pilot. He first joined a flying club in New Zealand when Lord Wakefield presented the country with six light aircraft. He left for Britain when it became clear that his home air force was too small. He cleared his Cape trip with his fiancé beforehand, and came to see the merit of flying with a less experienced pilot: she did not criticise his flying; she eased rather than aggravated his exhaustion; she helped manage their passage at airfields. The flight brought the usual avalanche of correspondence and invitations, which they handled for a while in two complimentary suites at the Grosvenor House Hotel. Lecture appearance fees earned Clouston £1,700 on top of the £1,200 which Lord Wakefield sent him to pay for the cost of hiring the two-seater Comet (before the flight he had declined to buy Clouston a 'coffin').

Aside from his signature moustache, the flight's most prominent reference to Empire was Clouston's winning the Aero Club's Britannia Trophy for the best performance in the British Empire in the air that year. Later in 1938 Clouston streaked to New Zealand and back in a shade less than eleven days. His 26,000 mile flight was Empire service of a kind: it got him home for the first time in eighteen years to his family farming roots. Along the way Empire service was crucial: Clouston and his male companion, the *Daily Express* air correspondent, had their Comet (*Australian Anniversary*) serviced at the Indian Aero Club at Cawnpore while they slept.[14]

After a route survey flight with his father in both directions across Africa in 1938, Alex Henshaw (twenty-seven) made a record-breaking return solo flight in a light aeroplane from England to Cape Town in

February 1939. Despite being told by two pilots whom he admired that he should use the *Imperial Airways* route as it was 'the only safe way across Africa to the Cape', he took the less well serviced but 400 miles shorter West African track which Amy Johnson had also used.

Like others in the late 1930s, Henshaw's flights had little imperial purpose but were imperial in geography, attitude and support. Once, at Alexandria, he gave short shrift to aviation regulations by taking off duplicitously. Fearing reprisal, he was relieved to see British battleships below him in the Mediterranean. Flying with his father between Kisumu and Khartoum, he elected not to make the regulation landings at Juba and Malakal because it would have delayed him and he was feeling ill. The Sudan's Governor-General commiserated and did not fling the rule book at him. On his solo flight, Henshaw articulated imperial feelings when noting that the Swakopmund aerodrome (in South West Africa) 'seemed strange and unreal, not like part of the British Empire'. Close by, however, at Walvis Bay the cheerful and willing Englishmen were less standoffish than the Germans. Henshaw's pride must have swelled on hearing an airport assistant remark of the civil ensign painted on the rudder of his aeroplane that 'there's no finer flag in the world than the Union Jack'.[15]

Having flown 6,377 miles in just under forty hours, Henshaw reached Cape Town in record time. Within a day he began the return leg, landing in England just under forty hours later. At the end of this marathon, he had to be lifted out of his plane. His physical condition did not persuade anyone that airline passengers could travel faster between England and the Cape. And, as he conceded, his flight should have failed several times.

The celebration of an astonishing personal achievement was not especially imperial at a time when war seemed close. Henshaw's aircraft was dismantled and reassembled for display at a Littlewoods retail store; maps and photographs were put on public display. There were the inevitable invitations, offers, interviews, propositions, broadcasts, lunches, dinners, lectures, first nights and sporting events.[16] But these were not used to promote Britain. There was no imperial clamour equivalent to that for two previous flying events.

Surmounting Everest

In 1924, the British mountaineer George Mallory failed in his assault on the world's highest mountain; a Briton would never be the first person on the roof of the world. It was a matter of honour that a Briton should be the first to fly over Mount Everest. The forecast made in 1919 that

Everest would 'succumb to the airman' was correct in all but one detail. The men over Everest were not pilots who would later tie British India to other 'civilised' centres.[17]

Mount Everest was never actually in the British Empire; the summit ridge is on the border of Nepal and Tibet. The entwining of Everest and Empire had more to do with the mountain being named after a surveyor-general of India than with either its location or the unassailability of the 29,000 foot peak. Everest, not Empire, epitomised the supreme challenge and the tragedy or triumph it brought.[18] So it was for a British airborne expedition buoyed by the British achievement of raising the aircraft altitude record to 44,000 feet in 1932.[19] Cockpit and engine designers had learned how to combat thin air and low temperature. The flying expedition was far superior to the hair-brained (and disapproved) plan by one Englishman to fly his Moth to India and crash land it as high as possible on Everest before ascending the summit on foot. His flight to India succeeded, but he disappeared during his climb.[20]

The 1933 Houston-Everest flight expedition acquired the first part of its name from its financial backer. Lady 'Fanny' Houston, an eccentric British patriot, was the widow of a millionaire Liverpool ship owner and Conservative Party politician. After her London physicians satisfied the Royal Court in Jersey that she was sane enough to handle her husband's legacy, her first gesture in the air was to fund Britain's challenge in the 1931 Schneider Cup air race. Ramsay MacDonald's Labour Government declined to support Britain's entry because, when the economic depression was biting hard, his pacifist administration did not want to lavish public funds on a sideshow that rewarded dashing daredevils and military aircraft designers. Aviation devotees and ardent nationalists were horrified to see Britain's prestige in the air being compromised by the Air Minister, Lord Amulree (Lord Thomson's successor), and his Under-Secretary, Frederick Montagu, both of them inexperienced officials. A suggestion made in 1930 to establish an Empire Air Fund had come to nothing. Houston intervened to save Britain from the feckless newspaper boys, shop assistants, free-lance hacks and magicians whom she deemed to be in charge of the nation's affairs.[21] Britain won the air race for a third time and, under the rules, won the prized Schneider trophy outright, snatching it from European rivals.[22]

Having once supported a British aviation venture, Lady Houston, Britain's second wealthiest woman, was persuaded to part with £15,000 to back the Everest expedition. Having himself been approached by Col John Buchan (a fellow M.P. from Scotland), Lord Clydesdale (a Scottish duke) approached Houston, one of his mother's friends. He was impressed by her intense feeling for Britain and Empire, but had no inkling that she saw the 'purely scientific expedition' as a way of impressing Indians with imperial might. He also 'had no thought of playing the heroic impersonation of British youth'. Soon he was mired in unwanted attention and controversy. Trying to convince his East Renfrew parliamentary constituency that he needed to take leave, Clydesdale exaggerated the flight's importance. His brief reference to the probable psychological effect of the flight on India drew 'wild protest' from the nationalistic Indian Congress Party, something which made him feel that Houston's belief in the political implications of the flight were sound after all.[23]

The Everest venture was planned meticulously and executed scientifically. Instigated by Maj. L. Blacker of the Indian Army while on home leave, assistance was received from the Royal Geographical Society, the College of Aeronautical Engineering, the Bristol aeroplane manufacturer and Shell.[24] For a time the committee managing the operation (most of them closely tied to India) worked out of a suite in the Grosvenor House Hotel, site of Johnson's tumultuous welcome. There was nothing secret about the project: a banner advertising the flight fluttered from the hotel roof. One of the people in charge of arrangements was William Forbes-Sempill (then still the Master of Sempill). A year before becoming the nineteenth holder of one of the most ancient peerages of Scotland, he was of impeccable social pedigree and was a particular favourite of Lady Houston's. The wartime flying hero, A/Cmdre P. Fellowes, who was in overall charge, organised the expedition in a mere thirteen months from start to finish. Aircraft, engines and photographic equipment were chosen, modified and tested. Pilots were trained. Two aircraft were crated and shipped to Karachi free of charge courtesy of the P&O shipping line. Members of the expedition who sailed out to India were ticketed at preferential rates. Additional material and several extra personnel (including Blacker) flew to Karachi on *Imperial Airways*. The airline would bask in an appreciative statement about reducing air travel to the fine art of lounging in an armchair while travelling gracefully through tumbled clouds and peeping at wonderful lands.

Working with the Royal Geographical Society, the India Office and the Air Ministry, Fellowes had to persuade the suspicious Nepalese that a forced landing would not be used as an excuse for Britain to occupy their country. No approach was made to Tibet's ruling theocracy. Lord Clydesdale, the chief pilot, noted that the team had also to contend with Nepalese anxiety about affronting the gods of Everest. Claiming to respect this 'vagary of mind', he placed it on a par with the irrational superstitions that Westerners (including even scientists) had about walking under ladders or lighting three cigarettes with one match.

The organisation of the air 'assault' on Everest, and its execution, eclipsed the ponderous fourth British Everest overland climbing expedition that set out in 1933. Speed, support and success were different, as was the social aspect. The programme included a nineteen-day flight to Karachi by three support aircraft, including Moths. At Cairo and Amman the expedition members used RAF facilities. Britain's overseas civil representatives hosted them at Baghdad and Bandar Abbas. At Baghdad, one of the team used an old Balliol friendship to get an audience with King Feisal. At Bushire, a British official used 'plenty of bluff and a little blarney' to settle the problem of missing travel documents. At Karachi, RAF men helped assemble the two high-altitude aircraft. The expedition pilots, who were based at the Sind Club, hired personal servants for the duration of their Indian visit and acquired their first 'close insight into the Oriental mind'. They were struck by what they identified as dishonesty.

Wherever the expedition landed in India *en route* to and from the base camp at Purnea, British civil and military agents, and 'loyal Indians', made them welcome. The Shell Company laid down special fuel dumps. At Delhi, the Viceroy and Capt. F. Tymms (Director of Civil Aviation in India) offered hospitality. At Jodhpur the team had the use of the Maharajah's aerodrome and house. All except three of his seventeen motorcars were made available. There were facilities for swimming, polo, pony racing, tennis and dining. At Purnea the Empire served the expedition team proudly. The Government Inspector of Works set up the camp. Townspeople organised a welcome under triumphal arches made of paper and waved airplanes made of tinsel. Eight miles east, at Lalbalu landing field, 'an army of coolies' pitched canvas hangars for the aircraft. As if going to a latter-day Jerusalem, some Indian potentates travelled to pay their respects. Their slate-grey elephants may have been lumbering, but their silk, silver, gold and gem coverings made them considerably more splendid than the flying machines opposite which they were photographed.

Indian spectators were not the only people to swell the numbers of Purnea for a month. The expedition's servants, caterer, telegraphers, doctor and film crew added to the invasion. From a commercial point of view this would have been welcome in a small town which was once the prosperous centre of the indigo trade, but which had suffered competition from synthetic dyes. The initial arrival of the flying British in the district was certainly exciting. A huge number of people gathered on the airfield, making landing difficult. Having failed to scatter them by low flying, the first pilot to arrive decided that 'fear must achieve

Figure 3 Two-speed Empire: Royal visit to Houston Everest team, Lalbalu, 1933.

what prudence evidently could not': he landed abruptly. On the ground, he harangued the pressing crowds in the forceful Hindustani 'which the British officer finds it necessary to cultivate in India'. Even so, schoolboys got close enough to poke pencils through the fabric of the tail plane.

In Purnea and Lalbalu joy was tinged with fear. Some local people suspected something more sinister than jaunting over Everest. One story circulated by anti-British agitators was that the real object was to fly to Turkestan and assist one of the warring factions in China. Clydesdale ascribed the 'grotesque' rumour to ignorance of both geography and aeronautics. Even so, he opposed the mid-air release of a smoke bomb that the camera-crews insisted on using to help them film the flight over Everest. After a trial explosion, rumour spread that bombing practice had started, and that an area round the aerodrome was to be evacuated. Trekking and disquiet lasted a few days. Belief that the British flying presence boded ill was compounded when a car belonging to the expedition team ran over and killed a goat. Compensation of the owner scarcely helped. It was fortuitous that the suspicious Indian villagers never found out that the aerial photographs which *Imperial Airways* flew to London went to the map-making section of the War Office.[25]

For all the help received on the ground, the success of the over-Everest flight itself owed nothing obvious to non-British skill and endurance. This was unlike the eventual climbing of Everest in 1953 when a Nepalese Sherpa stole the limelight. A 'generous Indian gentleman' did loan the Houston team his private light plane after a storm in Allahabad wrecked one of its own support aircraft, but his gesture did not upstage the mountain flyers. That the kindness should have been shown by Mr R. N. Chawla suggests that brotherhood in the sky could surmount a petty imperial snub.[26]

In London, *The Times* printed the first-ever overhead Himalayan photographs with considerable fanfare. The paper, which sent its aeronautical correspondent, E. Colston Shepherd, to cover the flight, scooped exclusive text, photograph and press syndication rights in a deal that guaranteed a £1,200 royalty payment to the expedition. Authorities were not seduced by the 'ballyhooston'. On the basis of existing topographic maps, the identities of the peaks in the three photographs were disputed. At Purnea a struggle was brewing over whether to attempt a second flight – one which would fly over the correct peak, but for which the aircraft were not insured. Writing to his principal, Shepherd revealed an atmosphere of anxiety, secrecy, suspicion and dissent surrounding careers, honoraria, news management, flying and release of photographs. In the event, a second flight was undertaken without Fellowes' authority through the expedient of confining him powerless to his sickbed a day longer than necessary. A rift opened between those whose interest in the expedition was aeronautical and scientific, and the publicity-seekers who treated it as a stunt on which to capitalise personally or commercially.[27]

Ignoring Fellowes' personal attempts to prevent the second flight, Day's hagiography of Lady Houston claimed that its success 'electrified the world'. Furthermore, it was said to have placed the prestige of Britain on a higher peak than ever. Specifically, 'it enhanced the power and might of the British Raj at a critical period when such a magnificent gesture was badly needed'. It also boosted the reputation of Britain's aeronautical industry. The Everest expedition was said to be the impetus behind the building of the Spitfire aircraft. Even after the heroic efforts of the RAF in the Second World War, it was possible to claim that it was in effect dowager Houston – a latter-day Boudica – who had saved Britain.[28]

Houston received one of twelve silver commemorative medals struck by *The Times*. Identical medals were awarded to nine members of the Flight. A *Times* presentation luncheon at the Grosvenor House Hotel in June 1933 was attended by almost 200 guests. There having been no aerial tragedy in the remote and inaccessible Himalayas,

Fellowes could explain humorously that pilots obeyed his orders to fly only within safe gliding distance of a flat space by treating clouds as flat spots. The British Empire Union hosted a celebratory dinner at the Mayfair Hotel in July. In a toast saturated with nationalism, Lord Danefort complimented the air team on proving wrong the 'melancholy growlers' who said the spirit of adventure was dead.[29]

The film footage shot by the *British Gaumont* cameramen during the expedition was edited for public presentation and prolonged the celebrations without attracting attention to either the disappointment of the first flight or the tensions surrounding the second. By the time it was screened in London in June 1934, Houston had become embroiled in another pro-British project. As new owner of the politically ultra-right-wing *Saturday Review*, a weekly paper on which Day was appointed editor, she filled it with jingoistic articles, poems and news, many of which she contributed herself. The paper was soon attacking the Government for its feebleness in abdicating the Empire. Lady Houston herself was feeling sidelined – the Government had not even bothered to thank her for her offer to contribute £200,000 to aerial defences for London. Indicative of her thoughts about the drift from cherished values, her paper's masthead at various times proclaimed that it was the only paper that dared to tell all the truth, and that it was 'written only for men and women who love their country'.

Lady Houston recalled her motives for funding the Everest flight on several occasions. She did so in the preface to the 1933 book *First over Everest* (reprinted three times in its first month of publication), in the opening sequences of the film *Wings over Everest*, and in her luncheon speech at the film launch. She used her newspaper to publicise her reasons fully after the short speech she would have given at *The Times* luncheon was paraphrased in her absence – the newspaper's proprietor was suspected of being embarrassed by a statement which contradicted his paper's editorial angle on India. Houston's explanation was as follows. A relation of hers who had arrived from India discovered that a neighbour there had been murdered. She told Lady Houston that violence was not unusual because agitators were permitted to preach treason. Many Indians assumed that Britons had lost courage. Houston hoped that 'some great deed of heroism' would rouse Indians and make them remember that although they were a different race, they were British subjects of a King who was also Emperor of India. 'What more can they want?', she asked haughtily and innocently. An imperial crusader, she hoped to stem Britain's waning prestige in the East, and to 'reinspire the people of India with confidence and faith in us, and in our good will and love towards them'.[30]

In the cringingly 'uneducated extravagance of her style'[31] which brought ridicule to the causes she defended – and which angered the Indian Congress Party – Houston patronisingly declared that crossing Everest would remind Indians about the advantages, privileges and loving British rule they had enjoyed: they had been fed during famine, nursed during plague and had benefited by administrative justice. Now it appeared Britons were on the verge of doing another favour for peripheral citizens of Empire: the organising secretary of the Everest flight noted that the team had found a high lake (duly named after Lady Houston) from which water might be piped to the plains of India, giving light and power to the millions of people 'ignorant of the treasure within their grasp'.[32]

Houston hoped that a successful flight would prove to Indians that English pluck and courage were alive. Others had used different techniques. For example, at the start of his American fund raising, before Mallory famously snapped that he wanted to climb Everest simply because it was there, he said that his 1924 expedition aimed 'to show that the spirit that built the British Empire is not yet dead'.[33] Equally, the aristocratic Everest expedition and the film were meant to help regenerate Britain and prove that 'the British Raj was still top dog, and still the Indian peoples' best friend'.[34]

It is unclear whether the film helped to spread a message about Britain's 'technological power and fitness to rule'.[35] In what he saw as a pervasive atmosphere of anti-imperialism, the *Saturday Review*'s film critic doubted whether the stirring, patriotic production would get a screening outside London or the Empire. Yet it was, in his opinion, something which audiences would prefer to the 'average flap-doodle of purely ephemeral sex-stuff and the rest, with American affinities'. As it happened, American judges in Hollywood rated the forty-one-minute film highly enough to award it an Oscar for best short subject in 1934. Lord Clydesdale thought the production an indiscriminate jumble that did no justice to either the expedition or the photography. A British reviewer scented pretentiousness and insincerity.[36] Any register of wider audience tastes was swept aside by an aviation event even more startling than the Everest expedition. It was one that affected the 'white' Empire's most British-inclined people living most distant from one another.

Racing to Australia

The England–Australia air race over 11,000 miles in October 1934 was held in conjunction with celebrations to mark the centenary of the founding of the Australian state of Victoria and its capital, Melbourne. Organised in London by the Royal Aero Club, the race was a logistical triumph, not least because of the international diplomacy involved in securing rights of over-flying and landing. Even for an air race, Colonial and Foreign Office contacts were indispensable. A seventy-three-year-old philanthropist, Sir Macpherson Robertson, a rags-to-riches confectioner, sponsored the event. MacRobertson, as he liked to be called, hoped that the winning aircraft and pilot would be 'Empire products'. Pilot and plane had two chances, one in the speed event and one in the handicap event.[37]

The MacRobertson race followed many previous dashes between Britain and Australia in the early 1930s. In 1931 there were eleven attempts on the speed record. There were seven attempts in 1932, and in the following year there were eleven more. Sheer national pride motivated some of the challengers. A mix of dare, opportunity and boredom motivated others, like Charles W. Scott. A London high-school dropout, Scott worked on a sugar plantation in British Guiana for eighteen months before joining the RAF and eventually washing up in Australia where, working for an airline, he was Amy Johnson's air escort from Darwin to Sydney. He knew thrill without glory, but set matters right when he began his flight from England to Australia on April Fool's Day in 1931. Albeit a Moth delivery for a client in Brisbane, he called his flight a purely sporting effort: he just thought it was time for an Englishman to hold the record. It would be the first of four times he held the speed record for a flight between England and Australia.[38] Speed was a drug long before the appearance of the eponymous amphetamine.

Scott arrived at his destination in eight-and-a-half days, shaving eighteen hours off Kingsford Smith's most recent record. Alongside advertisements proclaiming the merits of the spark plugs and the petroleum products used in Scott's plane, a *Times* editorial kept apart the motive and the meaning of record-breaking flights. In a fit of flag waving around concurrent flights to Australia (Scott) and Africa (Kidston), the paper praised the 'descendants of the old-time explorers and merchant adventurers whose daring and contempt of danger enlarged the bounds of this island [Britain] till it became the centre of a world-wide Empire'. Scott and Kidston were said to share the courage and toughness of their British predecessors, but to have embarked on a peaceful escapade shorn of any notion of conquest. They risked only their own lives and limbs. Their flights were not intended to add to the imperial possessions of Great Britain, but to bind its constituents closer together.[39]

Air speed records did not last long in the 1930s. In October 1933 alone, two records were set. Kingsford Smith returned to the fray and completed an England–Australia flight in seven days and two hours.

Clipping one-and-a-half days off Scott's time brought brief pleasure. In his own words, the 'stunt' took its toll – he 'went to pieces' in the Persian Gulf and spent a day in bed. At the end of his journey he felt run down and a bundle of nerves. In a radio broadcast from what he termed 'the new country' to 'the old country', Kingsford Smith noted that his British aeroplane outlasted him. Chasing fame was not much fun after all, and it could not have been pleasing to have one's best effort smashed almost instantly and reset at six days and seventeen hours. Regardless of how its poor-bloody-fliers felt, the British Empire could be said to be shrinking. Within fifteen years the flying time between England and Australia had been reduced from twenty-eight days to just less than seven. Between 1930 and 1933 the record was revised at the rate of roughly one day per year. With another great convulsion in 1934, the fastest time to the extremities of Empire would halve.[40]

Even before it started, the MacRobertson race caused a stir in England, Australia, Holland and America. The press and radio analysed, scrutinised and predicted. At stake was national prestige. In Britain, P. R. Burchall, then *Imperial Airways* Assistant General Manager, thought that the race would be a yardstick by which the efficiency of the country's commercial aviation would be judged. In his weekly *Field* magazine column, Lord Sempill declared it 'vitally essential' that Britain should win. A wartime pilot turned aeronautical commentator, Maj. Oliver Stewart, interpreted the race as 'a trial of strength between British aviation and the aviators of all other countries'. Certainly it would be profoundly ironic, as C. G. Grey remarked, if the part played by British enterprise in founding the state of Victoria had to be celebrated and remembered by the triumph of some foreign aircraft.[41]

A contingent, Schneider-like issue was whether failing to win would be worse than not even entering the race. Irrespective of imperial glory, in the austere financial climate of the time the British Government was reluctant to support any entry. Having different priorities, the aviation industry sided with the likes of Burchall, Sempill, Stewart and Grey. The aircraft constructor, de Havilland, noted bluntly that 'we can't stand by and let this race be won without any British effort'.[42] In great haste, and with only minimal aeronautical trials, his firm built three two-seater Comet racing aircraft for entry into the race. Like Britain's entries in the Schneider Cup races, these were another of the nation's 'shop-window' aircraft. They were built by what was little more than a cottage industry compared with American and German plants. It was five years since the United Steel Companies of Sheffield had bragged about the way British metal – not to say mettle – had helped win the Cup. Five years had also elapsed since *The Times*' financial editor naively wrote that just as the building of commercial ships and not warships had been the foundation of British shipbuilding, so 'thoughtful men feel that aircraft manufacturers too will find their surest progress in the development of commercial flying and not in work for the Air Ministry'.[43]

One Comet racer was crewed by O. Cathcart-Jones and K. Waller. Each pilot had at one time held the speed record for England–Australia and Australia–England flights. Tellingly, Cathcart-Jones originally hoped to race an American plane, but had difficulty finding a sponsor until he inveigled himself into the seat vacated by the Comet's ill owner. The pilots of a second Comet were C. W. A. Scott and T. Campbell Black. An English wartime pilot and social climber, Campbell Black was once Managing Director of Wilson Airways in Kenya. Both on aircraft delivery and other flights for Wilson's, he flew several people between London and Africa. His passenger on his seventh flight between East Africa and England was Maj. Jack Coats, the 'cotton millionaire'.[44] Aged thirty-three, he was also a pilot for Lord Furness and the Prince of Wales's safari in 1930. In 1931 he taught Beryl Markham to fly. He replaced Winifred Spooner as Lord Furness's permanent private pilot in 1932, and accompanied him on his annual shooting expedition in Kenya. In 1933, Lord Clydesdale crossed paths with Campbell Black at Cairo when on his fourteenth journey home by air from Nairobi.[45]

The husband-and-wife team, Jim Mollison and Amy Johnson, piloted a third Comet. Had their flight not been ended prematurely by engine seizure (and an alcohol-fuelled row) at Allahabad, the race may have assumed less of a heroic masculine tone. Flying together may have suppressed their competitiveness temporarily, but unfinished business left other challenges open. Their marriage lasted only a few more years before they separated in 1936. They divorced two years later. Women's magazines in Britain drew the moral that 'no good could come of attempts to play roles for which nature had not fitted women'.[46]

As race day neared, excitement rose on a tide of publicity. The *Daily Mail* printed a two-page colour map of the route. The compulsory stopping places between the start at Mildenhall military airfield and Melbourne fixed the mental map of the Empire airway: Baghdad, Allahabad, Singapore, Darwin, Charleville. The sensationalist *Daily Mirror* drummed up British hopes that the 'honour and glory' would go to a British pilot, a British machine and a British designer. Nationalism mixed with heroism, myth and legend. In the *News Chronicle*, the essayist Robert Lynd drew a parallel between the twentieth-century airmen and Hakluyt's voyagers. The contestants would 'face perils that are the very stuff of heroic literature – perils of storm, or wreckage, or forced landings in the swamp, or among savage tribes and wild beasts'. In its columns, the *Daily Express* (which, with a readership of 1.8 million had knocked the *Mail* from the top of the circulation tables) hinted at an event akin to a second coming. It reported that hundreds of tribesmen were riding day and night by horse and camel to the aerodromes at Aleppo and Baghdad. At first, the reporter wrote, suspicious tribesmen supposed Britain was flying out squadrons to chastise the 'Indians'. After their fears subsided, bazaars along the ancient trade route rippled with expectation.[47]

The race attracted sixty entries from fourteen countries. In the end, only twenty-one aircraft took off. Eight of these were from Britain, and three each from Australia, the United States and New Zealand. Statistically, Empire chances were good. Fifteen years since the Smiths first dashed across the Empire in twenty-seven days (spending ten per cent of the time – 135 hours – actually flying), the 1934 event was described as the most significant air race of all time. Some people regarded it as one of the great and unique voyages in history, rivalling Drake's circumnavigation of the globe. True to history, the participants proceeded on 'a knife-edge between life and death, between fantastic triumph and hideous failure, between luxury and insolvency'.[48] The competitors risked more than just personal hope and pride.

Scott and Campbell Black left Mildenhall after a lavish farewell banquet given by the owner of their plane, who happened also to own the Grosvenor House Hotel and had its name painted on the aircraft fuselage. Their Comet streaked to Melbourne in less than three days. The winning time was every bit as satisfying for Britons everywhere as Alcock and Brown's flight across the Atlantic in 1919. Lady Houston sent a telegram of congratulations that included the phrase 'rule Britannia'. A Melbourne radio announcer cackled that 'the breed triumphs again'. British genes, courage, tenacity and modesty had won through. Melbourne's Lord Mayor conveyed to Scott and Campbell Black the Empire's admiration, and Australia's homage. In his own tribute, MacRobertson declared that the epic flight confirmed the courage and endurance for which British airmen were famed. On BBC radio MacRobertson spoke of 'a stupendous achievement' deserving Empire praise. Other platitudinous references were made to the way aviation drew people of the world closer together, and how it might improve international understanding and enhance friendships.[49] Delivery of two current editions of *Aeroplane* magazine sneaked aboard the Comet certainly kept Melbourne up to date.

When news of the winning race time reached press offices, editors were quick off the mark. A gushing leader article in the *New York Times* supposed that the victors must have felt that the Empire had become more than a collection of widely separated colonies, and that they must 'tingle with a new and more satisfying sense of solidarity'. The British press succumbed to an orgy of patriotism about British pilots, the British sponsor, British designer, British engine-builders, and British suppliers of fuel, petrol, spark plugs, instruments and rubber tyres. Even the pilot's wristwatches, breakfast cereal, tomato juice, shoes, soles, socks and underwear were British.[50]

British through and through, it was a measure of progress in aviation that neither Scott nor Campbell Black was knighted. Technically they won the £10,000 first prize, but the Comet's owner claimed it and then sold the plane to the Air Ministry rather than to Campbell Black. Lord Wakefield gallantly wrote out a prize cheque himself. Even the medals from the Royal Aeronautical Society were silver rather than gold. The Britannia Trophy award for 1934 from the Royal Aero Club was standard. Homecomings were subdued in comparison with previous ones. When Cathcart-Jones and Waller (carrying some photographs for Scott's autobiography) were welcomed back to Britain a mere thirteen days after their departure, there were well wishers and newspaper reports, but no parades. It was enough to see and report patriotism flushed with pride by the return of a British aeroplane flagged with a Union Jack on its tail. Cathcart-Jones himself wrote proudly of having 'shown the flag through the out-stations of our Empire'.[51]

The MacRobertson race was followed microscopically. Sixty thousand people in England were reported to have watched the start. The same number saw it finish. They, and many thousands of Empire-wide newspaper readers in the interim, would have taken great pride in the result. Radio listeners tuned in to several broadcasts. One covered the start of the race; another featured a talk by William Courtenay. There was a description of the end of the race, and a recording from the winners. Several British cinema newsreel companies screened views of race preparations and of departing aircraft. *British Paramount* and *Pathe Gazette* gave top billing to the British victory. An animated sequence of still photographs of Scott and Campbell Black's arrival in Melbourne was shown in some cinemas. *British Paramount* scooped the Australian film footage that Cathcart-Jones and Waller flew back to England. BBC radio played a forty-minute reconstruction of the race.[52]

Behind the instantaneous joy, and the smugness, the race result was not a resounding British triumph. Warning bells were ringing about the standing of British aeronautics and *Imperial Airways*. The two aircraft that arrived second and third in Melbourne were standard production-line passenger aircraft built in the United States. At a time when *Imperial* owned just twenty-five aircraft, the first American airliner to arrive was a Douglas DC-2, one of a production batch of sixty. After stopping seventeen times *en route*, more than required by the race regulations, the plane finished the race less than twenty-four hours after Scott and Campbell Black, and won the handicap prize. *The Times* salvaged some pride by likening the plane rudely to 'a van-horse' which had 'taken a day off from its round to run second in the Grand National'. The plane was certainly less sleek than the Comet. It was also heavier in design (modern, all-metal) and load: it carried a crew of four, three fare-paying passengers (two Dutch bankers and a female aviation journalist from Germany) and 25,000 letters.[53] All were emptied out to lighten the aircraft for take-off in poor conditions late in the flight in Australia.

The DC-2's load handicap gave the result aeronautical interest; the badging of the aircraft added aviation significance. From the point of view of *Imperial Airways* it was humbling that the DC-2 was entered by the Dutch airline, KLM, which had simply wanted to show how quickly and comfortably passengers could fly on its regular commercial service to and from the East Indies.[54] Before a British commercial air service had even begun between Australia and Britain, a rival national airline had demonstrated a level of service that *Imperial Airways* could not hope to beat. KLM's business-like approach, and low profile, were epitomised by the two Dutch pilots of the DC-2. In his 1938 BBC talk 'Four Famous Air Races', Maj. Stewart recalled that visitors to Mildenhall approached the two men dressed in unobtrusive blue uniforms seeking directions to the refreshment tent.[55]

Another standard American civilian airliner landed in Melbourne only hours after the DC-2. This Boeing was another aeroplane that Britain should have built. Its own long-distance airliners were dinosaurs by comparison. The American plane beat the Comet which Cathcart-Jones and Waller flew into fourth place; it outperformed the third Comet that Johnson and Mollison withdrew from the race at Allahabad. The Boeing did not fly in any airline colours but its success was no less visible. Stewart saw the issues sharply. In the *Saturday Review* he tartly amplified the point he had been making since the spring of 1934 about the stagnation of British aviation. Were it not for one heaven-sent Comet, he said, 'one might weep with rage and mortification at the backwardness of British aviation'. He argued that although Britain may have won the greatest air race in history, it had scarcely started the even greater race toward worldwide commercial (and military) air superiority.

Writing to *The Times*, Lt-Col J. T. C. Moore-Brabazon (M.P., and the first Englishman to fly in England) shared Stewart's despair. He denounced the 'deplorable technical performance' of British civil aviation. Shamefully, in his view, an American airliner overtook even a fast

RAF scout plane in the race. As if that were not mortifying enough, his version of the calamity was that the stunned RAF pilot saw half-a-dozen women passengers glancing nonchalantly out of the DC-2's cabin windows while 'powdering their noses'.[56] In a panic of national self-abasement, the number of passengers, and the number who were female, were exaggerated. Humiliation had to be complete. Was it in the sky – previously masculine space – that the familiar world order would begin to crumble? And what then of Empire?

Defenders of the faith spoke passionately but did little, as the aeronautical industry found out. Even de Havilland had difficulty capitalising on its race success and manufacturing reputation. The firm wrote to the Air Ministry three times hoping to recoup some of the Comet's development costs. It also sought financial assistance to develop an aircraft and engine suitable for sale in world markets, and for flying on overseas routes where foreign airlines competed. De Havilland's observation about how far Britain lagged behind in the development of high-speed commercial aircraft fell on deaf ears. The appeal for assistance to maintain British prestige went unheeded. Other manufacturers – Armstrong-Whitworth, Short and Handley Page – also realised that *Imperial Airways* managed its fleet and its operating targets too well to necessitate change. The airline could only be more enterprising outside the limits it was set. The large-capacity, economical airliners which were prescribed for its all-stations, combined mail and passenger services, were just not suitable for any other airline. In an exclusively domestic market, the costs of British civil aircraft development were spread over the small number of airliners ordered by the official Empire airline.

Great Britain won the MacRobertson race, but failed to turn the cross-Empire event into a bigger victory. The race for aircraft orders was won by the American aeronautical industry. de Havilland turned away from trying to build a fast airliner of moderate size for export, and concentrated on building successful light aircraft. By comparison, the American airline industry used liberal state assistance to meet domestic and foreign airline orders for fast, metal monoplanes for a booming passenger market. Britain was told to be content. Holland and Germany, meanwhile, doled out subsidies to their external air services flying the fast, modern, imported American airliners.[57]

The MacRobertson race was indeed a mixed blessing for both British domestic and Empire aviation. On the positive side, the name de Havilland shone brightly for a time. Furthermore, the event did give colour, absorbing interest and gaiety to a world that was just recovering from a disastrous slump and was soon to slither into an even more calamitous conflict. The race helped to compact the British Empire and also gave definition to its parts, at least for outsiders. Amy

Johnson noted this in her 1935 lecture to the Women's Engineering Society. Whereas in 1930 Allahabad was 'merely a piece of land called an aerodrome', she said, when it was chosen as a control point for the England–Australia air race it quickly became a fully fledged air station. She would not have known about behind-the-scenes politicking which might have kept Allahabad off the map: the Maharajah of Jodhpur petitioned the Air Minister directly to persuade the race organisers to use his own aerodrome.[58]

Set against the positive aspects of the race were the unpleasant public argument about Britain's place in world aviation and the despondency that overtook the hamstrung aeronautical industry. No such block choked Britain's shipbuilding or railway locomotive industry when the Empire was served only by overseas and overland transport. Beyond Britain, Australia's isolation and vulnerability in the Pacific basin may have been altered *vis-à-vis* Britain, but so too had the continent's proximity to Asia altered. Would aeronautical advance weaken rather than strengthen Empire? It was plain that technological progress could not be kept from non-Empire countries if American aircraft would always be available for purchase.

The disturbances which aviation might in future create in prevailing arrangements at an international scale were already being preceded by smaller tremors. Fierce competition in the air was a recipe for disharmony. The race sparked an international incident involving the entries from the United States. It created suspicion in Germany, irritated the Turks (who refused to rescind their ban on Kingsford Smith flying over Turkey because of an indiscretion), and provoked an undignified quarrel about prizes between the Royal Aero Club and the Royal Aeronautical Society.[59]

On a personal level, the race wrecked one marriage, narrowed the friendships of Sir Charles Kingsford Smith and temporarily ruined his reputation. He was branded a coward for not joining the race, despite getting financial help from MacRobertson. Some thought it was his just reward for having chosen to fly an American plane. In fact, he had been unable to procure a suitable British aircraft. He faced the same problem later when trying to purchase aircraft for his trans-Tasman airline linking New Zealand and Australia. Kingsford Smith avenged some of his critics, partially restored his status and flew the Australian flag high by completing his second trans-Pacific flight. He left Australia twelve hours before the MacRobertson race started.[60] Not least, the events surrounding the race embittered Cathcart-Jones. In 1935, six months after the finish and 'the biggest banquet of the year' in celebration at plush Claridges Hotel, he scribbled a furious denunciation of the British Government for its conceit in declaring the race a victory for England when it had made no attempt to finance the building of race aircraft.[61] Down from the clouds, that year he raced his Lagonda luxury marque to first across the line in the Monte Carlo rally.

Appetites for Empire air racing were truly jaded by the time C. Scott and G. Guthrie won the 1936 air race between London and Johannesburg. Indeed, long before, in November 1934, Britain's aviation authorities had responded coolly to South Africa's proposal that Sir Abe Bailey put up £25,000 for an air race from London to Cape Town to celebrate the coincidence of the Union's and the King's twenty-fifth anniversaries in 1935.[62] Campbell Black made two attempts on the Cape record that year. In the end, the wealthy South African industrialist, T. Schlesinger, sponsored a race in 1936 in connection with the City of Gold's golden jubilee and the Empire Exhibition there in September. The race rules stipulated that aircraft had to be of British make and that Cairo was the only compulsory stop. There were few entries because British aeronautics was preoccupied with orders for rearmament. One entrant, Campbell Black, was killed shortly before the race in a runway accident at Speke airport, Liverpool. In something of a fiasco, Scott and Guthrie's aircraft was the sole finisher out of nine starters. Their flight took two days.

Arthur Clouston's race took longer and was a cameo of imperial aid. At Khartoum, the RAF fitted his plane with a set of replacement engine pistons obtained from *Imperial Airways'* stocks. South of Khartoum, where Clouston made an emergency landing, he was helped by the Scottish manager of a sisal plantation. After crashing near Bulawayo, he was driven to the city by a 'railway executive'. At Johannesburg the penniless pilot was accommodated by Sir George Albu, the mining magnate, who gave him funds for a first-class steamer ticket back to England.[63]

Other race experiences (including one fatal crash) were a poor endorsement for flying, not least in comparison with the *Imperial Airways* display in the Empire Exhibition hall devoted to shipping, travel and aircraft. The airline displayed intact aircraft models and safe-flying aids such as wireless and navigation equipment. In the transport-themed display in the British pavilion at the Exhibition, the centrepiece was a huge world map showing the Empire's sea lanes and air routes. Relays of light bulbs wired in sequence fast-flashed the approximate routes of *Imperial* aircraft. The light plane which Amy Mollison, (nee Johnson) flew on a record double journey between Britain and Cape Town in May 1936 hung from the ceiling above the map.[64] The doomed Schlesinger race was disconnected from the Exhibition and relegated to an unfortunate blip in progress toward connecting the Empire better.

Notes

1 P. Zander, '(Right) wings over Everest: high adventure, high technology and high nationalism on the roof of the world, 1932–1934', *Twentieth Century British History*, 21 (2010), 300–29.
2 The most recent account is S. W. McKay, *Mildenhall to Melbourne: the World's Greatest Air Race* (Berkhamsted, 2009).
3 Cecil Day Lewis's poem 'A Time to Dance' recalls Parer and McIntosh's flight at due length.
4 J. H. L. B. Hinkler, 'My flight to Australia', *Journal of the Royal Aeronautical Society*, 33 (1929), 1025–33.
5 *The Times* (23 February, 13 March 1928); *Punch* (29 February 1928), p. 212; T. Gwynn-Jones, *Farther and Faster: Aviation's Adventuring Years, 1909–1939* (Washington D.C., 1991), p. 229; R. D. Mackenzie, *Solo: the Bert Hinkler Story* (Sydney, 1963).
6 Peggy, the daughter of a wealthy London businessman and property developer, was actually aged twenty-four. Some London papers played up her poor-little-rich-girl image and commented on the evening gown she packed for her Cape Town reception. Luff, *Mollison*, p. 133.
7 Jones, *Time*, pp. 160–1; *Star* (Johannesburg) (2 November 1931), p. 10; *Sunday Express* (8 November 1931).
8 *The Times* (7, 10 April 1931); *Airways* (April 1928).
9 www.gettyimages.com/detail/82145205/Hulton-Archive (accessed October 2010); Jones, *Time*, pp. 156–7.
10 *The Times* (13 March 1928).
11 *Spectator* (17 March 1928), p. 405.
12 *Popular Flying* (January 1936), pp. 531–5, 575; (May 1936), pp. 82–3, 95, 108; *Flight* (14 November 1935), p. 509; (2 January 1936), p. 6; (12 March, 1936), p. 278.
13 *Sphere* (October–December 1937); *Flight* (18, 25 November 1937); Jones, *Time*, pp. 190–8.
14 A. E. Clouston, *The Dangerous Skies* (London, 1954).
15 Henshaw, *Flight*, pp. 111, 158–9, 168, 270.
16 Henshaw, *Flight*, pp. 195, 281, 284–5.
17 H. Woodhouse, 'High-altitude flying in relation to exploration', *Geographical Review*, 7 (1919), 149–58.
18 W. Unsworth, *Everest* (Yeovil, 1989).
19 Lady Bailey and Mrs de Havilland set a world altitude record of 17,289 feet in 1927.
20 McKay, *de Havilland*, p. 181.
21 J. W. Day, *Lady Houston DBE: the Woman Who Won the War* (London, 1958); *Airways* (October 1930), pp. 465–6.
22 A 1934 map book of *Imperial*'s European air routes contained an illustration by the prominent graphic artist, Edward Bawden, depicting the soaring figure of Britannia holding the Schneider Cup.
23 D. Douglas-Hamilton and D. F. McIntyre, *The Pilots' Book of Everest* (London, 1936), p. 9.
24 Zander, '(Right) wings over Everest'.
25 P. F. M. Fellowes, et al., *First over Everest: the Houston-Mount Everest Expedition* (London, 1933); Douglas-Hamilton and McIntyre, *Pilots' Book of Everest*, pp. 80, 106.
26 Fellowes, et al., *First over Everest*; P. F. M. Fellowes, 'The Houston-Everest flight', *Journal of the Royal Aeronautical Society*, 38 (1934), 689–704; L. V. S. Blacker, 'The aerial conquest of Everest', *National Geographic*, 64 (1933), 128–62; B. R. Lewis, 'First flight over Everest', *Aviation Quarterly*, 6 (1980), 168–79; P. H. Hansen, 'Tenzing's two wrist-watches: the conquest of Everest and late imperial culture in Britain 1921–1953', *Past and Present*, 157 (1997), 159–77.
27 *New Statesman and Nation* (13 May 1933); News International Archive (London), Mount Everest Flight files: Shepherd to Deakin, 20 April 1933.
28 Day, *Lady Houston*, p. 207.
29 *The Times* (24 April 1933); (2 June 1933), p. 8; (5 July 1933), p. 11.

30 Fellowes, et al., *First over Everest*, p. ix; *Saturday Review* (9 December 1933); (9 June 1934), p. 656; (9 January 1937), p. 42.
31 R. Graves and A. Hodge, *The Long Week-End: a Social History of Great Britain, 1918–1939* (London, 1940), p. 270.
32 P. T. Etherington, 'Into the stratosphere: the airway of the future', *Chambers's Journal* (1935), 17–19.
33 Unsworth, *Everest*; *London Review of Books* (7 May 1998), p. 8.
34 Day, *Lady Houston*, p. 205.
35 Zander, '(Right) wings over Everest'.
36 *Saturday Review* (9 June 1934), p. 656; (16 May 1936), p. 631; Douglas-Hamilton and McIntyre, *Pilots' Book of Everest*; *Sight and Sound*, 3 (1934), p. 74.
37 A. Swinson, *The Great Air Race: England–Australia, 1934* (London, 1968), p. 7; NA, AVIA 2/2280.
38 C. W. A. Scott, *Scott's Book* (London, 1934); McKay, *de Havilland*.
39 *The Times* (11 April 1931), p. 11.
40 *The Times* (12 October 1933), p. 12; BL, NSA 1CL0067162; Swinson, *Great Air Race*, pp. 24, 131.
41 *Spectator* (19 October 1934), pp. 561–2; *Field* (18 August 1934), p. 412; *Saturday Review* (22 September 1934), p. 142; Swinson, *Great Air Race*, p. 44.
42 C. M. Sharp, *D.H. - A History of de Havilland* (Shrewsbury, 1982), p. 138.
43 C. Barnett, *The Collapse of British Power* (London, 1972); *The Times* (9 September 1929), pp. 19–20.
44 *East Africa* (24 October 1929; 4, 11 September 1930; 15 January, 12 March 1931). Campbell Black's passengers on the 1929 aircraft delivery flight for Wilson's included an aero engine mechanic, the son of Nairobi's town planner and Capt. H. C. Druett, editorial secretary of *East Africa*.
45 Douglas-Hamilton and McIntyre, *Pilots' Book of Everest*, p. 37.
46 M. Pugh, *Women and the Women's Movement in Britain, 1914–1959* (London, 1992), p. 76.
47 Swinson, *Great Air Race*, pp. 108–9.
48 Scott, *Scott's Book*; Gwynn-Jones, *Farther*, p. 251; Swinson, *Great Air Race*, pp. 19, 125.
49 *Saturday Review* (20 October 1934), p. 291; BL, NSA 1CL0067301; Swinson, *Great Air Race*, pp. 138, 191.
50 Swinson, *Great Air Race*, pp. 139–40.
51 A. S. Jackson, *Old Pilots, Bold Pilots* (Gillingham, 1998); *The Times* (3, 6 November 1934); O. Cathcart-Jones, *Aviation Memoirs* (London, 1934), p. 265.
52 BBC Programme Records, 1933–34; *BBC Annual* (1935); *Listener* (31 October 1934); BNIS, *British Paramount* (25 October, 5 November 1934); *Pathe Gazette* (15, 18, 22, 25 October 1934).
53 *The Times* (24 October 1934); *Aeroplane* (24 October 1934).
54 Gwynn-Jones, *Farther*; H. C. Kavelaars, 'The DC-2, the Dutch, and the 1934 Melbourne race', *American Aviation Historical Society Journal*, 29 (1984), 256–74.
55 BBC Written Archives (Caversham) (WAC), microfilm text, 31 January 1939.
56 *Saturday Review* (20 October 1934), p. 427; *The Times* (6 November 1934), p. 10.
57 Sharp, *D.H. - A History of de Havilland*, pp. 157–9.
58 Swinson, *Great Air Race*, p. 196; NA, AVIA 2/2280.
59 Swinson, *Great Air Race*.
60 B. J. Hurren, *Fellowship of the Air* (London, 1951); *Aeroplane* (19 December 1934); B. Sheil and C. Simpson, *Caesar of the Skies* (London, 1937).
61 *Saturday Review* (20 April 1935), p. 491.
62 *South Africa* (23 November 1934), p. 226.
63 Clouston, *Dangerous Skies*.
64 *Daily Telegraph* (14 September 1936); *Aeroplane* (23, 30 September 1936); *Star* (Johannesburg) (17 September 1936).

4

Imperial encounters

Far from being disembodied machinists, private pilots who dashed about the Empire were social creatures. Their individual characters and dispositions were unique, of course, but they were all moulded in late imperial times and traditions. Men and women pilots projected their acquired attitudes while away from Britain, and refracted what they saw through personal lenses. Theirs was a new generation of Empire travellers. They were new commentators on British contact with foreign places and people.

Ascension and condescension

Not all private pilots were given to writing and public speaking, and one can only surmise that most were struck at least by the physical markers that they used for navigation. On the trans-Africa airway it would have been difficult not to notice the Nile, the pyramids, Abu Simbel, the Rift Valley escarpment, the Great Lakes and the Victoria Falls. These landmarks would not necessarily make every pilot reflect on the transience of civilisations, the progress of technology, the power of nature, and the ardour of imperialists like Livingstone, Stanley, Burton and Speke. Yet even the least cerebral man-in-a-kite would have been amused by the pleasures and curiosities of the happy-go-lucky colonial life encountered while flying across Africa. Games of polo and golf, country club dinner-dances and balls at government houses were arranged for flying visitors. In reply they hoisted up local officials and dignitaries for aerial 'flips'. They saw town mayors and district commissioners dressing in garb befitting officers and gentlemen, and watched aircraft being guarded by sentries of the King's African Rifles marching as if outside Buckingham Palace.

Solo pilots may have felt the social burden less than flight leaders. A member of the 27,000 mile RAF 'flag wagging' flight in 1927/28 in the Middle and Far East and Australia recalled that the duty of flight captain was similar to that of the captain of a touring MCC cricket team. Playing ambassador, diplomat, public relations officer and film star, the top man had to be polite to foreign statesmen, crashing bores and chattering women. He needed to answer innumerable silly questions and to parry sensation-seeking journalists. He was expected to endure endless entertainment, speak wittily at short notice, admire local attractions and shrug off parochial prejudices.[1]

A/Cmdre C. Samson, who led the RAF's annual flight from Cairo to Cape Town and back in 1927, was certainly struck by the social challenges of imperial flying. He and his men regularly got less than four hours' sleep. Well prepared, he selected his flying officers not just for their good manners, athleticism, smart appearance and cheery, lively dispositions, but also because they danced well, could participate in any festivity, and could play polo, golf and tennis. Regarding flying itself, Samson reflected quite deeply. He noted, for instance, how travel at height offered new perspectives: 'from the air you are able to blank out the objects of modern civilisation, and gaze on the old structures unspoilt by a modern foreground'. Instead of developing this thought about a new imperial gaze, he ruined it by adding that a lofty passage meant one could avoid being disturbed 'by the constant horde of pestering guides, camel-men, donkey-boys and other touts'. Like earlier transient travellers, pilots' views were formed quickly during the aeronautical equivalent of a whistle-stop tour. A 'propeller-stop' outlook on Africa would have had less weight than the views of sedentary missionaries, administrators and anthropologists, but it was transmitted more quickly.

Samson reported 'unbounded interest' in the RAF flight among White settlers (he was told that it fortified their patriotism to the 'Mother Colony'), but he was sceptical about the wider 'civilising' role of his official flight. He judged that its impact (and presumably also that of private flights) was not altogether a good thing. It may have been close to heresy to write about the 'evils of civilisation', but his judgement was the antithesis of political radicalism. The British presence, Samson thought, spoiled the servility of rural Africans especially. Insisting that men 'straight from the bush' made the best servants – 'the natives at Abercorn always used to run with our kit' – he wondered how long it would be before civilisation provided them with trade unions to stop them working so hard. Africans, so he judged, were more contented living under 'semi-civilisation' than in a locality which had reached a high stage of so-called advancement.[2] Aviation, it seemed, did not ennoble what was regarded as 'savage'.

Maj. Sir Ralph Furse, in charge of recruiting at the Colonial Office (and himself an *Imperial Airways* passenger on the Juba–Khartoum leg of his 1935 trip through Africa), had his own thoughts about flying in Africa. Whether it was good for Africans or useful for White people who interacted with them, he agreed that geographical isolation prevented Africans becoming impertinent and preserved countryside courtesies. Bad roads, no telegraph (and presumably no air links) eased White people's dealings with Africans.[3] C. G. Grey, arguably the most influential aviation journalist in the Empire, sympathised with such views. In his *Aeroplane* magazine he circulated his own rabid thoughts. Drawing a comparison with 'low-class Irish', he dismissed the Egyptian fellahin (peasants) as 'savages spoiled by a veneer of civilisation', and declared that 'a saucy educated nigger may be a nasty proposition'.[4] The new technology of flight had failed to elevate the spirit of one of its prime activists and opinion-formers.

Speaking one's mind was a rather insensitive and un-British thing to do. Events should be left to speak for themselves, and language used to make points by omission. After the completion in 1926 of an RAF flight from Cairo to Nigeria, Hoare told Parliament that the 6,300 mile journey had been an unqualified flying success, as well as a great imperial success. He based his assessment on the reception given the flight: 'hundreds of thousands of natives' gathered to see the aircraft; pilots were addressed as 'bird masters'. The Emirs (whose old-fashioned attitude and dress Hoare took as a reflection on British modernity and cultural superiority) were amazed. Among those who were taken up for joy rides, one remarked that 'the English had brought trains, then motor cars, and then the wind train, and there was only Allah left to see'.

The flattery might have drawn a wry smile from the Empire's high flyers. An enigmatic chuckle might have greeted the observation (first made by the flight commander, and in a reader's letter in a May 1926 edition of the *East African Standard*) that during a flight of four RAF aircraft over Kenya, a tear-stained herder referred to the strange birds as 'Mathewie, Markie, Lukeie and Johnie'. That year the Swahili word for bird (*ndege*) entered the vernacular as the term for aeroplane. Kenyan chiefs and headmen who inspected an *Imperial Airways* liner at Nairobi coined the 'long bird' (*indegi m'refu*) variation in the early 1930s. Cobham was named *bwana nkubu ya ndege* ('great master of the bird') during his 1926 flight through East Africa. His mechanic and photographer were a lesser species known as 'small masters'.[5]

Overnighting in the bush among 'primitive' and seemingly uncomprehending people put Western aviation achievements into perspective. Cultural stereotyping could and did follow. Simplistic cultural-anthropological thinking moderated it on occasion. Interviewed in London, the normally taciturn van Ryneveld remarked that when he and Brand passed low over hillside villages in southern Sudan, Uganda, East Africa and Northern Rhodesia on the first ever trans-Africa flight, chickens were more visible than humans. It appeared that villagers had 'got the wind up' and had instinctively taken cover. At one stop, however (at Shirati on the shores of Lake Victoria), the people 'regarded us as gods, and brought peace offerings of milk, food and fruit, while an old chief came out with goats and sheep and wanted to kill them for us. We made use of the old chap however. We told him that the machine would not go until it had its morning bath, and he, straightaway, set his forty-three wives to clean and oil it up'.[6] Such tasks were probably gestured. A decade after his own concurrent African aerial adventure in 1920, it was Dr P. C. Mitchell's view that there was no need to speak an African language: 'all African natives accept the name George and understand such amenities as a shake of the hand, cigarettes, grins and gestures'.[7]

Sir Alan Cobham, who could be dismissive, recorded that when he visited Malakal during his 1925/26 African flight, Sudan's Shilluk people took no interest in his aeroplane, seeing it as 'just another example of the white man's madness'. Yet, with some insight, he noted that other imported technology (water tanks, pipes, taps) fascinated the Sudanese because it related directly to their lives.[8] An RAF commander on a three-year tour of duty in the Sudan sensed rather less. In 1934 he noted that 'all over Africa the native looks upon the white man as mad, particularly the white man who flies'.[9] A Royal Navy pilot who flew himself from Britain to South Africa sensed contempt from Sudan's 'black savages' who, unlike the indignant giraffes, were unaffected by the passing of his mechanical bird'.[10]

Madness and badness are close companions, and the RAF commander's book did not shy from portraying Africans as uncivilised, lazy, murderous, immoral and superstitious. His view was that flight should be used to pacify, rule and control, 'not to destroy the native, but to coerce him'. Recalling Shakespeare's Cymbeline, he wished for a horse with wings as a new way of 'enforcing our will on naughty natives'. Naughtiness wore different guises, and the young commander's text also included a rare reflection on sexual temptations awaiting people who took to earth the freedom of the skies. Africa, it transpired, was also diseased. The RAF commander, who had graduated from being President of the aeronautical society at Cambridge in 1925, warned against venereal infection from the 'lunacy' of exercising too great personal liberty. In remarks intended for all connected with flying in Africa – RAF personnel, airline crews and ground staff, and private pilots – the commander urged men to decline offers of hospitality from tribal chiefs, and to resist the 'dusky charmers' and 'winsome virgins' who might disturb one's sleep. Simultaneously polarising and stereotyping, he insisted that 'the attachment and attraction of the fair for the dark, the dull for the witty, the clever for the clown, white for black, must not be made an excuse'.[11]

Mutual curiosity and pleasure probably lessened first in places where regular flying became established. Elsewhere, first flights made a notable impact. The Air Survey Company light aircraft chartered by a British research scientist, Dr Humphreys, created an ambiguous stir during his visit to Uganda in 1931. Humphreys made a reconnaissance flight before trekking overland with porters in the Ruwenzori range (the famed 'Mountains of the Moon') and their brace of high peaks named after Stanley, Baker and Speke. He observed that reactions to aircraft were exaggerated, or at least varied. Africans at Msindi were blasé about aeroplanes; they regarded them just as European contrivances. In their ignorance they had to be stopped from walking into invisibly spinning propellers. During taxiing it was sometimes necessary to threaten African spectators, whereupon they disappeared 'like chaff down wind'. In contrast, when Humphreys' plane glided down toward a remote highland village, the inhabitants rushed out of their huts and prostrated themselves in terror. The pilot's instinctive reaction was to open the throttle and leave the scene quickly. The roar was hardly placatory.[12] Similarly, when first sighting aircraft used by the Anglo-Persian Oil Company for aerial surveying in Papua New Guinea, locals would disappear into the bush: 'one tribe, in its consternation, collected its pigs for the final feast and prepared for the Day of Judgement'. In time, Papuans regained their confidence and sometimes walked long distances to see an aeroplane close up.[13]

Among Africans, passive acceptance, even stupefaction, accompanied repetitive flying visits. In 1933, the *Imperial Airways* house magazine excerpted from the London *Evening News* the smirking observation that 'the modern Ugandan native takes little notice of this new wonder which weekly passes over his head'. Using gendered pronouns correctly in the case of the pilots, but probably not the onlookers, the paper continued that the apparition which first evoked blank astonishment when real men emerged from the huge bird quickly ceased to be a matter of comment: 'it passed into his accepted scheme of things and now interests him no more than the passing motor-car, the steam roller on the township roads, or the hot tap in his master's bathroom'. It was condescending to add that, for Africans, it was enough that the aeroplane had been 'brought by the Europeans – all is thereby explained'.[14]

Familiar aircraft shapes may have lost their spectator novelty quite quickly, but new arrivals would continue to turn heads. When the first *Imperial Airways* flying boat arrived at Dar es Salaam in 1937, the whole town turned out, according to one journalist. If nothing else, crowds at airfields could be useful for navigation. Sometime in the early 1930s when approaching Dodoma on his first aerial tour of inspection, Sir Stewart Symes, the Tanganyikan Governor, was surprised at how well the landing ground was marked out. He soon saw that the long rectangular lines were in fact 'thousands of natives in white cotton robes, with black faces upturned to greet the arriving plane'.[15] The reader is left to suppose that the gathering was not orchestrated.

Master or madman?

First-time and once-off aeroplane landings in remote corners of Empire created the greatest stir. The wealthy Irish engineer-aviator, Francis McClean, first stirred Arabs during a three-month flight in a seaplane along the Nile in 1914. Egyptians (who for years called all aircraft 'McCleans') were apparently terrified by the sight of the bi-plane. After one of thirteen engine breakdowns, two 'native carpenters' who were required to do some repairs refused to board his plane (although the engine had been removed) in case it flew away with them. Curiosity overcame fear in the case of fellahin who waded out towards the bi-plane whenever it landed.[16] These initial encounters could infuriate pilots who then gave vent to deeply ingrained attitudes. Sir Alan Cobham revealed the clash of attitudes and cultures in his 1927 book about his Australia flight. The upsetting death of his photographer would not have helped his mood. Beyond Baghdad, a bullet fired from the ground killed the cameraman. It transpired that an Arab gazelle hunter had on several occasions had his quarry stampeded by low-flying RAF aircraft, and when again a target was startled the annoyed hunter fired upwards in anger. Eight RAF mechanics stationed in Basra acted as pallbearers at the burial. A Union Jack was draped over the coffin.[17]

Following an accidental collision with a launch belonging to the British Consul at Bandar Abbas, Cobham wrote contemptuously about the 'poor, half-witted, cross-bred, nondescript coolies of the Persian Gulf', and about natives with 'a very stupid temperament'. His English aide appeared to be less flustered by the locals. Were there class-concealments in Cobham's observation that his replacement aircraft mechanic, a toolmaker by trade, was 'a wonderful fellow with natives'? Leaving Bahawalpur, the mechanic trod on the hands of an over-eager 'native assistant' to stop him clinging onto the aeroplane floats during take-off. Elsewhere, Cobham repelled boatloads of Indian spectators by squirting them with water from syringes; they objected strongly to having their best white clothes spoiled. Was this the renowned British sense of fair play? What had become of the idea that international aviation would enhance understanding of 'the other fellow's point of view'?

However annoyingly, local people did actually help Cobham, but he disparaged the gangs that helped him to rope, anchor, tow and haul his plane across eddies and through reeds. He condemned a Burmese steamboat captain and his 'stupid' assistant at Rangoon, neither of whom had ever seen a map before. After taking off again, he recorded that 'the whole crew stood aghast and are probably talking to-day of those weird people who dropped out of the skies . . . upset all their arrangements, caused them to stop their engines, cut their best manila rope, and left them again without a thankyou'. Cobham took for granted that local people were unoccupied and were there solely to help. He bullied them into action: 'natives were cursed at freely, caught hold of and placed in position, and by practical demonstration shown how either to push or hold. Europeans were commandeered to control the natives.' On another occasion, Cobham's assistant rushed along a line of people standing in loincloths under umbrellas in the rain. He seized their umbrellas and flung them aside. The 'coolies' stood stupefied; 'Europeans' looked on convulsed with laughter.[18]

Relations between aircraft pilot and local people were not necessarily conflictual. There is no sign of tension in the photograph that shows scantily clad men carrying Cobham shoulder-high on a makeshift litter away from his seaplane. Porterage was scarcely necessary: the seaplane is beached in shallow water which barely covers the bearers' ankles; only a few yards away, a local official stands on dry shore in long trousers. Cobham was at no risk of drowning, and could easily have clambered out of his plane without even wetting his shoes. The 'tribesmen' (as the caption to a 1981 photo print termed them)[19] may have received Cobham spontaneously as a king, but it is more likely that they were instructed to fetch him. The moment, actual or re-created, was a golden photo opportunity. The event itself was a splendid colonial cameo, a tale in and of black and white, frozen for posterity by someone with a commemorative consciousness. The great man's stage entrance from the wings was a performance oozing symbolism. The sea remained a link between the old world and the new; the custom of carrying delicate White people through the surf persisted. The landing set off the primitive against the modern. It elevated dressed White men above semi-naked Black men, the flier above the carrier, the leading actor above the stagehands. The angle of the camera lens has the seaplane towering over the porters: foreign technology is superior to the rickety indigenous litter. The photograph's themes are subservience and dominance.

When Ross and Keith Smith made their epic England–Australia flight in 1919, they too were amused by an incident (at Timor) when 'swarms' of locals gathered to watch their departure. Some who took up a vantage point on a fence tumbled backwards in the propeller slip-stream. The crowd of ponies and people 'made a glorious mix-up at which we all laughed heartily'. Earlier in their flight the Smiths were helped at Singora (Siam) where one of the three resident Englishmen had 'imposed upon the simple native minds that the devil was going to arrive in a flying chariot to take charge over all the convicts there'. Actually, the Dutch authorities arranged for 200 prisoners in leg-irons to clear tree-stumps from an inadequately flattened take-off path. Light repairs were made to parts of the Smiths' Vimy in a local rice mill. Labourers worked after-hours pedalling a human-powered lathe. When they declined to continue, another gang of four was 'secured'. They trudged for thirty minutes before demanding more money. Smith paid up, but fifteen minutes later the men struck again. The foreman who was summoned struck out altogether differently.[20]

Some two or three hundred Javanese came to the Smiths' aid at Surabaya, another of their twenty-eight stops. It was not a spontaneous gesture: the harbour engineer 'collected a horde of coolies' who spent thirteen hours hauling the Vimy out of the mire and laying down a 300-yard-long runway of reed mats. One commentator interpreted the 'most commendable' eagerness with which people stripped their houses of roof and floor coverings as a piece of the jigsaw of air deeds that held the Empire. Other pieces were slotted in later. In 1922, Maj. Blake's failed round-the-world flight was assisted just outside Delhi, where he made a forced landing. The public works department 'lent' hundreds of Indians to fill in ditches, cut through earth banks and chop down hedges and trees so that the aeroplane could be hauled to a road. There was evident compliance, if not enthusiasm.[21]

Smith liked to report other facets of foreignness. His observations (and the language in which he cast them) told of naive, primitive people. Natives at Rangoon hurried to the racecourse as soon as the local press reported the Smiths' departure from London; they were expected to land at any moment. Nearer the actual time of arrival at Rangoon, many local people took food and bedding to the racecourse, where they intended holding a festival for the duration of the Vimy's stay. At Bima, in the Dutch East Indies, villagers scampered in all directions when the Smiths landed. The crew aroused intense curiosity. Eyes peered furtively at them through every gap in their allotted bungalow. One onlooker overstepped the bounds of decency by rousing Smith. The Australian pilot woke to a movement outside which he described in animalistic terms as 'prowling', yet he had no hesitation writing that he waited until the man was opposite the doorway before firing a pistol which 'put him to a screaming . . . terror-stricken flight'. With the possible exception of this frightened man, at sunrise the natives overcame their shyness and 'swarmed' round the aircraft with gifts of coconuts. Smith presumed that they thought the Vimy was a very thirsty sort of bird.[22] Was it inconceivable that honouring and blessing were within the compass of other humans and would take unfamiliar forms?

Alan Cobham's book about his flight round Africa in 1927/28 discloses more of the contrasting and ambivalent attitudes toward people on whom overseas flying depended in some measure. He admired and was grateful to expatriates. Among these were the District Commissioners at Berber and Mongalla, the Kenyan Governor at Entebbe and the British Consul at Libreville. Cobham's party enjoyed hospitality from the Portuguese at Port Alexander and Luanda; from the Commissioner at Takoradi; from the Governors at Freetown and Abidjan. The Commissioner and Senior Marine Officer at Bonny, who travelled especially from Port Harcourt, made themselves useful by getting local people to fetch bars of soap to plug a hole in the flying boat. The Commissioner also arranged for six Africans to work in three pairs all night to pump water out of the hull. Cobham kept one of his men on watch throughout, just as he kept a wary eye on the Africans who contrived to 'swarm' all over the flying boat when it was lifted by government dockyard crane out of the water for hull cleaning and painting at Lagos. The weight of a huge crowd threatened to sink the jetty. Cobham walked defensively round his flying boat with a wet paintbrush, dabbing anyone who got too close. Some scampered and struggled to retreat; others took it as a joke. A few tumbled off the jetty into the water twenty feet below. The only recorded casualty was on the streets of Lagos, where a crowd spilled out to see what was happening when the flying boat arrived. An African was killed by 'casting his body under a motor car in his excitement'.

Cobham found Africans either helpful or annoying. They were seldom beyond remark. The tinsmith in the 'native quarter' at Khartoum who helped Cobham did so with equipment that looked 400 years old. Cobham's inference was that Africa was backward, not that simple tools were ageless or that the modernity of flight was overstated. One can read between the lines of Cobham's text about who was to blame for the damaged and leaking cans of petrol that ended up on the lakeside at Butiaba (Uganda) after an overland journey from Mombasa. Porters certainly did not know that burying the cans in sand to reduce evaporation would corrode the containers – the sand had high soda content.[23]

Later, during his shorter 1931 survey flight, Cobham must have come close to tears when he discovered that 'fuel cans' laid down at

Lake Kivu were full of water. One African porter, having punctured a petrol can, emptied out all the others to lighten his load, reasoning that he could replace the colourless liquid at his journey's end at the Lake. Other African episodes were more successful, albeit qualified. Cobham reported that natives were very good to him and his crew when at Shambe village *en route* to Juba. There the Shell petroleum agent (an African trader) provided blankets and mattresses from his store. Iron bedsteads were borrowed from villagers. After a good night's sleep, Cobham wrote, 'it was fortunate that we were up early as the "livestock" in the native beds began to get busy at dawn'.[24]

On the Kivu expedition a motorboat piloted by an African twice rammed the floats of Cobham's seaplane.[25] During his 1927/28 flight, Cobham was alarmed twice when river craft at Mwanza came threateningly near the thin skin of the flying boat and its floats. One close shave involved an African temporarily in charge of a river launch; another involved Africans paddling an old pontoon. The sight of an armada of twenty-four Buganda war canoes with spiked metal prows bearing down on the flying boat on Lake Victoria was also unnerving. The ensuing welcome by hysterical oarsmen ended when the Commissioner blew a whistle.

Despite these two disconcerting experiences, when canoes circled Cobham's moored flying boat at Bonny, Nigeria, his team enjoyed the welcoming spectacle. Not so at Abidjan, where Africans aboard another old tub were warned off vigorously by the representative of the Elder Dempster shipping line. Cobham was unsure whether the man's shouts in French were about going to Hades, being shot or being flayed alive. He had to do his own yelling at Port Etienne, where sailing boats came so close that their masts became entangled with the propellers. At Bathurst an African messenger in a motorboat did actually collide with the flying boat. The repair took a day.

Elder Dempster contacts assisted Cobham in several ways. Its European representatives organised 'cases of drink and hundreds of sandwiches' at Lagos. They arranged for fifty Africans to scrape the hull of the repaired flying boat for nine hours once it arrived at Freetown. When Cobham left his flying boat and some crew on a remote lagoon in West Africa for four weeks while he and his wife went off in search of supplies and a place where he could cable Rolls Royce cryptically for an engine replacement, they fetched up at Grand Bassam where they were given the use of an Elder Dempster house. On that occasion Cobham's entire party also relied heavily on the canoes, paddles and muscles of local villagers. After twenty canoes towed the stricken flying boat to safe water, the two Cobhams set off for a trading post in a canoe with sixteen paddlers on a six-hour journey. Food was sent back to the flying boat the same way. Three canoes lashed side by side were used to get the new engine in its packing crate back to the lagoon.[26]

As if re-creating the dependence of early maritime Empires on coastal surfboats and paddlers, and of early continental Empires on indigenous porters, the flights of Cobham and the Smiths depended on local labour and sometimes on indigenous transport. Their contacts with colonised people reveal a common chasm of understanding between pilot and one kind of earthling. During the Houston-Everest expedition, an aircrew that strayed off-course was twice surrounded by a surging mob of thousands of shouting and gesticulating Indians. After one emergency landing, the 'brown, struggling mass of men, women and children inextricably mixed up with dogs, cows and donkeys' were kept at bay by revving the engine. After a second landing, Indian police 'waded in with sticks and staves' to shield crew and aircraft from a curious crowd estimated at 20,000.[27]

An account of a private flight that came to earth unexpectedly between Cairo and Aswan in 1934 tells of the pilot and his Irish passenger beating back the crowd that gathered round the plane 'like flies'. Four years later, a private pilot on his way to taking up a job as a flying constructor in Kenya felt similarly anxious when, after taxiing his plane up the main street of Atbara (Sudan) in order to attract attention to his presence and need for fuel, he was surrounded by 'swarming' Africans who chattered unintelligibly and turned his plane into 'flypaper'. In the 1934 incident, fretting that their plane would be reduced to matchwood, pilot and passenger floored two dark-skinned spectators and hauled away two others by the scruff of their necks. After intervention and hospitality from the local chief, the visitors received assistance from people who they labelled 'boneheaded'. A gang helped them take off. Once airborne, and presuming some shared comprehension of a farewell orbit, the pilot gestured his gratitude by circling over the chief's hut.[28] The abused villagers might have feared a second visitation.

The mixture of fear, bewilderment and gratitude is evident in the account that Victor Smith gave of his experiences in the Sahara after a petrol leak forced him to land his Moth during his effort in 1932 to break the Cape–Britain solo flying record. Interviewed in London, the nineteen-year-old spoke about spending twenty-four hours alone before a near-naked man appeared: 'he looked more like a beast'. Smith gave him two francs and a note to take to 'some white people'. The messenger turned out to be a slave of Tuaregs, whom the newspaper described as 'a fierce nomad tribe who prowl about French West Africa'. Smith spent two nights alone, scared partly by the roar of lions whose tracks he later saw 200 yards away. He sustained himself on two chocolate bars, biscuits and oranges. A water pool forty-five minutes away was foul. In time three robed Tuareg chiefs arrived on horseback together with some (robed) tribesmen and (naked) slaves. As they approached, Smith switched on the navigation lights of his aeroplane. The men backed away in awe; Smith reckoned they thought he was a god. Mutual fear subsided. The men appeared to be perfectly friendly and offered goat's milk in skins. Smith was unable to drink what the newspaper's sub-editor termed an 'unpalatable present'. One man tried to touch the aeroplane but the chief pushed him back. 'During the whole of the time I was with these strange people no one ever touched the aeroplane', Smith wrote.

The Tuaregs took Smith to their thorn-surrounded encampment two or three miles away. They erected a pole-and-cloth shelter for him and assigned a man ('slave') to sleep at his feet. They were really 'extraordinarily kind' for having housed me, he wrote. After three days the messenger returned with two native French soldiers. The commandant at Dori sent 'white man's food' and sixteen gallons of petrol. As he flew away to Gao his hosts gave him a parting salute like a ceremonial farewell. 'I had been with them for five days and had been treated by them in their own way as if I were an honoured guest from another world.'[29]

The attitudes that were part and parcel of flying imperially were also apparent during the reconnaissance flight made by Alex Henshaw and his father in 1938. After making a forced landing in Niger some distance from Niamey aerodrome, Alex (like Smith) asked an African to take him 'to the nearest white man'. The route on which they started out made Alex suspicious, and he feared a spear in his back. He was relieved to be picked up by a passing French motorist who, after translating, explained that the African had been avoiding ground where there were snakes and scorpions. Henshaw senior also experienced fear while waiting at the downed aircraft overnight. Human heads that popped up in the bush and then shrank back made him feel trapped, outnumbered and largely helpless despite the ever-present pistol. Making the aircraft engine roar was always a good way of making 'inquisitive natives' scatter.[30]

Clouston opted for a different tactic after his forced landing south of Khartoum in 1934. He too felt the pressure of motionless 'statues' who stared at him. He tried speaking in English, French and German. After a time, when the 'jet black and completely nude men with disproportionately long legs like swamp birds' still had not hurled their spears at him, he approached the man whose neck ropes Clouston thought marked him out as the leader. 'Bwana!' (Chief!) he roared. 'Bwana, Bwana!' the men chanted. Status settled, the 'chattering, high-spirited tribe' trailed the two men across three miles of swamp to the nearest expatriate.[31]

Figure 4 Alex Henshaw about to set off 'to find some white people' after a forced landing in Niger, 1938.

Financial reward or compensation always paid off embarrassment, shame and gratitude. Alex Henshaw decided to reward his African guide well when he got to Niamey. He wanted to do the same after Africans whom the French called in helped to push the aeroplane along a road to the aerodrome. Avoiding obstacles was 'an absurdly easy job with so much manpower available'. If there was any tree or bush overhanging the track, 'an overwhelming force of natives would completely erase it in no time'. The Frenchmen advised Alex to pay the Africans less than he proposed: generosity would cause 'endless mischief for weeks'.[32]

Winged goddesses

Before Johnson and Batten, the most prominent British women who flew across the Empire were the Duchess of Bedford, Lady Bailey and Lady Heath. The Duchess's relations with the people she encountered on her flights to India and Africa – her thoughts and behaviour – would have been modulated by the absence of the pressure that bore down on pilots who were chasing records. She was certainly quite restrained in her social commentary. The unclothed women at Juba were 'finer' than the men; African villages looked like 'a clump of mushrooms enclosed in a fairy ring of cacti to keep out wild beasts'; chickens, not Africans, fled from aircraft; it was 'Press fever' that had Africans wearing charms to ward off aeronautical 'evil'. The Duchess paid more attention to the 'indescribably awful' hotel at Dodoma than to the uniformed Askari guard of three.

Anxiety affected most private fliers; gratitude quite quickly overcame embarrassment. A violent thunderstorm between Dodoma and Juba obliged the Duchess's party to land on the *Imperial Airways* emergency airfield at Nimule. Pilot's and passengers' accounts differ, but a 'weird collection' of approximately fifty nearly-naked Africans arrived at the site. The Duchess recalled that the 'very black but very cheerful' natives dressed in girdles and beads were as amused by her appearance as she was by theirs. Using a torch to keep predators at bay, they guided the fliers to a shelter for 'stray aeronauts' and prepared food. In a rousing send-off the next morning, they wore as decoration the lids of cigarette tins given them the previous night.[33]

Not unlike the Duchess, at the conclusion of her two trans-Africa flights Mrs Westenra gave more prominence to mosquitoes than African men. Recommending flying to anybody who wanted to see a great deal of Africa in a very short time, she noted suffering only from mosquitoes 'which got into the cabin when the machine had to be left out in the open for the night'. She advised protection against malaria, and a good sleeping-bag. The 'small pistol' she suggested was more about fending off animals than men. Otherwise, there was 'no need to be armed', she said. The one White person she and Capt. Macintosh met was an American who had crossed the Sahara in a motor car: he 'looked much more dangerous than any of the people he was likely to meet'.[34] Thousands of miles away, the leader of a 10,000 mile aerial survey of Central Australia in 1930 had tried to distinguish between malevolence and ignorance among indigenous people seeing aircraft for the first time. His view was that terms such as 'hostile natives' and 'savages' which he had seen in print 'should not be applied to the natives'. It was a qualified rebuke: 'handled in the right way and kept out of one's camp they are certainly a help, not a danger or menace to travellers'.[35]

The danger some fliers felt was less about themselves than their aircraft. As Cobham had found, there could be hazardous encounters between boats and visiting aircraft. In 1931, Delphine Reynolds and Flt/Lt Pudney reported that their seaplane had been rammed by a river steamer whose Gambian pilot was asleep. Repair materials had to be shipped from London. Elsewhere in West Africa they dealt with Africans mishandling a mooring buoy and with over-enthusiastic locals hampering seaplane beaching. The London *Times* had anticipated unreliable and unpleasant natives.[36]

Aeronautically innocent spectators also showed unrestrained

curiosity on land, as Mrs Keith Miller found *en route* from London to the Cape in 1935. Departing from Croydon in mid January in a two-seater bi-plane, her trans-Sahara route from Tangier and Colomb-Bechar was via Gao and Niamey to Cotonou on the Atlantic coast. After flying for nine hours on the last leg, her plane ran out of fuel and she elected to land on a road. Twice, Africans 'swarmed' out of the bush on her approach. On a third occasion she landed. Swerving to avoid four curious spectators, she hit a tree and damaged the aeroplane. By the time she crawled out 'practically every Native in Dahomey had arrived on the scene. They were all almost naked, and spoke in an unintelligible language, but appeared to be friendly.' Nauseated by the 'smell of their moist black bodies' she eventually made them understand she wanted help carrying everything out of the aeroplane up to the village. 'And a strange procession started forth. Everyone wanted to carry something and every article was promptly placed on their heads.' Mrs Miller spent three weeks in the village in the company of two English women.[37] Her account of the accident might have been the sort of claim about which aviation insurers became wary, and which would have compounded pilot neuroses.

Jean Batten had mostly favourable impressions after a forced landing in Baluchistan during her first (unsuccessful) attempt to fly from England to Australia in 1933. Surprised locals ('tribesmen') helped lift her plane out of the mud; she posted two villagers as guards; others helped her on a nine-hour camel journey to reach a settlement from where she could telegraph Karachi for a new propeller. During a truck trip to Karachi to collect it, local people built causeways and dug ramps across five rivers. The return trip to her grounded aeroplane involved six camels, a pack of young boys running in front with torches, and panting men trotting alongside taking turns shouldering the propeller. 'These were the days when the British ruled India and a white skin was a guarantee of service and dutiful obedience', wrote Batten's biographer. She capitalised quickly on her privileged status, one made special by her descent from the heavens. Her queenly white flying suit, her radiance and her aura of superiority positioned her as a flying goddess and helped command unquestioning subservience from the Baluchi people. 'The glamorous and demanding aviator had dominated Baluchi peasant life for miles around for over forty-eight hours. Nothing quite like it had ever happened before and people would speak of it for generations to come.'[38] Batten's money-reward to the headman and his people was not necessarily the gesture that fixed the episode in popular memory.

Lady Bailey's comments about the indigenous people in the places where she landed are not as revealing as Lady Heath's, discussed later, being confined mostly to standard observations about eager, noisy crowds surrounding her Moth. One vignette about landing on a street in a Nigerian village describes how the antics of the crowd, and the men who were trying to whip the throng into order, turned into a pantomime. At a Malian village she politely packed into her cockpit a chief's gift of two live chickens and six eggs that she didn't want. At small aerodromes 'a swarm of natives' were apt to appear. At Tehumba, she waited for two hours 'with hundreds of excited Africans' until the local administrator arrived, bringing a guard to safeguard the aeroplane for the night. Her view was that 'the natives are perfectly friendly but intensely curious'. She certainly appreciated their assistance. Once, on the way to Mopti (Mali) when she landed near a village settlement to ask directions (and bent the axle owing to the restricted size of the ground and not being able to land into wind), the village chief found her a horse to go to the nearest centre. Returning from Mopti with a mechanic, she presented the chief with some white cloth as a present. The next day he appeared in a white dress sown overnight.[39]

Unusually, Lady Bailey commented on frustrating encounters with Europeans. The British authorities in Cairo prevented anyone (male or female) flying alone across the Sahara. The hostile desert compounded tribal instability that had first led to the killing of a Sudan District Commissioner in December 1927, and then to RAF retaliatory bombing. By the time Heath started her second (abortive) London–Cape Town return flight over the Christmas/New Year period in 1932/33 in an attempt to break the record set by Amy Johnson, the Trans-Sahara [road transport] Company required a deposit payment to cover any search-and-rescue expenses. For its part, the Colonial Office showed irritation about the time and cost involved in liaising with its overseas representatives by telegram when private pilots went missing.

An indication of the strain inflicted on British out-stations by search-and-rescue activities in the early days of British private imperial flying emerges from the Persian Gulf region. There, Mrs Victor Bruce's 1930 flight seemed doomed after oil pressure failure forced her to land in the wild. It was three days before she was rescued. Reported missing, a search party (which included an *Imperial Airways* ground engineer) found her in southern Iran 'after a safe but nervous time with local Baluchis'. A tone of annoyance crept into an official report by way of observation that Mrs Bruce had only six week's flying experience.[40]

An incident in November 1934 may not have been the next one to involve a private aeroplane. This time, the offender was Lady Blanche Douglas. After a forced landing sixty miles south of Bushire, a local man was sent to ask for help. As soon as he received the message, the British Resident at Bushire, Lt-Col T. C. Fowle, dispatched four soldiers to guard the plane and its occupants. The officers moved the two

English people into a headman's house from their burrows under the aircraft wings where they were sheltering from abominable weather. At the first opportunity, two official cars set out carrying food and a mechanic. Heavy rain obliged the rescue party to abandon the cars and walk nine miles. The downed light aircraft was reached after a two-day trek. After a broken fuel pipe was repaired, the pilot tried to take off for Bushire. After five attempts to lift off from soggy ground, Lady Douglas was removed to lighten the plane. Her overland trip to Bushire with her rescuers involved crossing a swollen river, and rescue by the pilot who brought a third car containing spare parts to repair the original two which had broken down. Lady Douglas and pilot stayed at the Bushire Residency for ten days before flying onwards. She was also hosted in the Residency on her return flight from India.[41]

By coincidence, in 1928 both Bailey and Heath were fortunate to arrange a Sahara escort in the person of Flt/Lt R. Bentley. The three converged while Bentley was completing the return segment of a London–Cape Town flight for which he would be awarded the Royal Aero Club's Britannia trophy. Experience merged with masculine gallantry (and payment from Lady Heath) when the northbound Bentley and his new wife spent a few days of their honeymoon flying first alongside Heath, and then, doubling back, alongside Bailey.

Continued application of the solo-flying prohibition obliged Lady Heath to plan her flight back to London on a route west of British-controlled Sudan. In the event, she was prevented from crossing the French Sahara by administrative indifference to her requests for fuel supplies, and was obliged to track the westerly coastal route instead. Lady Bailey admired French air initiatives in West Africa, and was generally grateful for hospitality in French-speaking Africa. Nevertheless, she jotted down some tart remarks about French colonialism. They put in perspective her subsequent plea for British air services in the continent. The French, she declared, 'hardly do a solitary thing to open up or assist the natives in their Colonies'. She added that the French 'ought not to be allowed to grab another handful of land anywhere, until they have the decency to do some good in what they have got'.[42] Flying visits over colonial interiors could reveal quickly rather a lot that evaded most foreign observers.

Lady Heath (who, for a time, had taught Lady Bailey to fly) started out from Cape Town for London in February 1928. Before beginning her 6,150 mile flight in a light plane, she gave lectures and flying exhibitions, and helped to arrange an air race in South Africa. She distributed several hundred Air League pamphlets; 3,000 people attended her lectures. Taking some 700 people aloft, she raised £1,200 for South African flying clubs.[43] On her way north, she focused more single-mindedly on her own project. Overall, her flying time to Cairo was seventy-two hours. She spent twice that time on routine engine maintenance. Leisure was typically imperial. During the day Heath would sometimes play tennis. At Mwanza she went on a buffalo-shoot. Flying past, she delivered the day's *Bulawayo Chronicle* to the mine manager at Wankie colliery. Her evening diary included sipping sundowners, dancing at the Pretoria Country Club, partying at the District Commissioner's home at Abercorn, Northern Rhodesia, and dancing at the British Residency in Cairo.

Now and again, Heath's trajectory north meant that she followed the footsteps of imperial notables, reliving and renewing the spirit of exploration. Towns named after Salisbury and Livingstone lay on her route. Flying over Tanganyika she passed Tabora, where Stanley had stayed sixty years previously, prior to meeting Livingstone. Beforehand, she flew over the Matopos hills in Southern Rhodesia, where Cecil Rhodes, architect of Empire in southern Africa, was entombed. From those heights he had once cast an acquisitive eye over what he considered was Britain's continental 'hinterland' reaching to Cairo. Heath's reflective mood over the Matopos may have been different had she known that her shipboard companion on the sea voyage out to South Africa was not in fact the man he claimed to be: Rhodes had no son.

Heath's 1928 trans-Africa flight was not just self-indulgent. Her material observations were useful. Among others, she saw a market for airfreight. Her list of likely commodities reads like a latter-day inventory from the Promised Land: ivory, spices, antimony, silver, pearls, furs, ostrich plumes, gold, platinum, radium. 'All these riches that now reach the outside world by slow costly journeys on naked backs and river cargo-boats will some day take wings along the trails my little machine helped to blaze', she wrote. She was most proud of her protest (and the subsequent parliamentary question) about the absence of a chain of radio stations in British Africa comparable to the efficient networks in the French and Italian colonies. The single telegraph-line running 500 miles south of Abercorn operated for only three months of the year. At other times, storms, ants, giraffe, and the crafty makers of copper necklaces and bangles, decommissioned it.

Like van Ryneveld's interview, and like Cobham's record of his pioneering flights into Asia and Africa, Heath's pen did more than record; it inscribed imperialism. She found herself 'a winged goddess' among awe-struck local 'tribesmen'. Their grasp of trains and motor cars related to their labour building railways and roads, but, she argued, the noise of a flying machine and glint of sun on burnished metal and oiled fabric caught Africans off guard. Her comment was least valid in places where Africans built not just railways and roads but also airfields. But even there, people would be surprised when the first 'great carriage that flies like a hawk' swooped down, disgorging its load of magicians. In the Transkei district of South Africa, an elderly 'witch doctor' at Tsolo sent a bead headdress with his respectful salutations to the White woman 'whose magic exceeded his'. Her powers did not extend to topping up aviation fuel which, hauled 100 miles by ox wagon, was unaffordable. Heath flew on wearing her gift. It was 'a beautiful salaam from a proud spirit, tribute from a primitive race to one that has evolved a more scientific magic'. In Northern Rhodesia, by contrast, an elderly Swahili man showed respect to Heath but was not at all amazed by her aeroplane. 'But, Memsahib', he asked, with a courtly bow, 'was it not made to fly?'

There are more insights into the imperial mind in Heath's report about the help she received from 'friendly, trusting Kaffirs' after she made a forced landing in the vicinity of Bulawayo. Suffering debilitating sunstroke, she instructed local African women ('my black girls') how to lash down her plane for the night, after which they carried her to their hutted harem, where they fed and washed her. She slept on her fur coat, using her silver-fitted dressing-case as a pillow. The visitor left her 'willing maids' abruptly the following morning, rescued this time by passing settlers. It must have been in their company, rather than on the basis of a brief semi-conscious stay in Matabeleland, that Heath learned about 'that peaceful native reserve'. Later, at the Anglo-American Company's copper mines near Broken Hill, she indulged in diving her plane down upon crowds of Swahilis and 'Kaffirs' who screamed and danced 'with childish glee'. Heath admitted that her dangerous but thrilling stunt was 'highly reprehensible'.[44]

Notes

1 G. E. Livock, 'The Far East flight of the Southampton flying boats, 1927/28', *Journal of the Royal Aeronautical Society*, 75 (1971), 327–35; G. E. Livock, *To the Ends of the Air* (London, 1973), pp. 140–41.

2 NA, AIR 2/299 (777467/27); C. R. Samson, *A Flight from Cairo to Cape Town and Back* (London, 1931), pp. 8, 165.

3 R. Furse, *Aucuparius: Recollections of a Recruiting Officer* (London, 1962), p. 263.

4 *Aeroplane* (4 January 1933), p. 54; (5 May 1937), p. 576.

5 *Hansard* (Commons) (25 February 1926), cols 773–4; *Airways* (October 1926); *Air Travel and Commercial Air Transport* (January 1934), p. 42; *Daily Mail* (15 January 1926), p. 10; NA, AIR 2/299 (695962/26).

6 *Flight* (6 May 1920), p. 504.

7 *Cape Times* (8 March 1930), p. 7.

8 A. J. Cobham, *A Time to Fly* (London, 1978), p. 98.

9 E. L. H. Williams, *Something New Out of Africa* (London, 1934), p. 181.

10 C. W. Byas, '"Swiftly" to the Cape', *Blackwood's Magazine* (1932), 742–67.

11 Williams, *Something*, p. 201.

12 N. Humphreys, 'Ruwenzori: flights and further exploration', *Geographical Journal*, 82 (1933), 481–514.

13 *The Times* (25 January 1928), p. 13.
14 *Imperial Airways Gazette* (December 1933).
15 *The Times* (23 April 1937), p. 16; S. Symes, *Tour of Duty* (London, 1946), p. 166.
16 *Flight* (11 April 1914), p. 389; Williams, *Something*, p. 186; G. Butt, 'An Irishman's diary', *Irish Times* (24 May 2011).
17 C. Cruddas, *In Cobham's Company: Fifty Years of Flight Refuelling Ltd* (Wimbourne, 1994).
18 A. J. Cobham, *Australia and Back* (London, 1926), pp. 35, 60, 63, 114; *Airways* (November 1926); *Journal of the Royal Aeronautical Society*, 31 (1927), 12.
19 O. E. Allen, *The Airline Builders* (Alexandria, VA, 1981), p. 28.
20 BL, OIOC, Mss.Eur. F150/6 & 7; R. M. Smith, 'From London to Australia by aeroplane', *National Geographic* (1921), 229–339.
21 L. E. O. Charlton, *Deeds that Held the Empire: by Air* (London, 1940); W. T. Blake, 'The first world flight attempt', *Discovery*, 4 (1923), 2–6.
22 R. M. Smith, *14,000 Miles through the Air* (London, 1922).
23 Cobham, *Twenty*; Cobham, *A Time to Fly*.
24 NA, DR 9/39: Kivu Flight Report, p. 107.
25 NA, DR 9/39: Kivu Flight Report, p. 85.
26 Cobham, 'My flight'; Cruddas, *Cobham's Company*.
27 Fellowes, et al., *First over Everest*, pp. 200–3.
28 C. K. J. Coggle, 'A forced landing', *Chambers's Journal* (1934), 593–8; *Flying* (21 May 1938), p. 29.
29 *Manchester Guardian* (25 November 1932), p. 9; V. Smith, *Open Cockpit over Africa* (Cape Town, 1992).
30 Henshaw, *Flight*, pp. 133–4.
31 Clouston, *Dangerous Skies*, p. 59.
32 Henshaw, *Flight*, pp. 137–8.
33 Gore, *Mary*, pp. 266, 268–9, 280–4; Curtis, *Winged*, pp. 79, 87–9.
34 *The Times* (18 January 1932), p. 9.
35 D. Mackay, 'The Mackay aerial survey expedition, Central Australia, May–June 1930', *Geographical Journal*, 84 (1934), 511–14.
36 *The Times* (2, 24, 30 March, 6 April 1931).
37 *Popular Flying* (August 1935), pp. 268–71.
38 Mackersey, *Jean*, pp. 98, 101.
39 Falloon, *Throttle*, pp. 117, 133, 142–5; *The Times* (29 January 1929), p. 15.
40 Bruce, *Nine*; PDPG (October 1930).
41 *The Times* (3 January 1935); PDPG (November, December 1934; January 1935).
42 NA, CO 323/1182/13; Heath and Murray, *Woman*, p. 214; Falloon, *Throttle*, p. 139.
43 Cluett, et al. (eds), *Croydon*; *Air* (May 1928).
44 S. M. Heath, 'My Cape Town to London flight', *Asia* (February 1930), 81–7; 'Sunstruck in the air', *Asia* (March 1930), 200–207; 'On to Cairo', *Asia* (April 1930), 284–90.

PART II

Commercial flying

5

'PAX' Britannica

. . . governors, generals, bishops of far-flung dioceses, heads of great banks and commercial firms with power equal to that of governments, and names that, east of Suez, were household words . . . its passenger list was a microcosm of colonial and overseas society, high and low.[1]

John Morley's recollection about imperial travel is about the P&O shipping line, not *Imperial Airways*. Yet – excepting for the every last word – it might as well have been about the airline. Such, anyway, is the folklore about who used Britain's designated overseas carrier to fly about the Empire: wealthy people were the passengers (acronym 'PAX' in current airline parlance) on scheduled civil aircraft services in the 1930s on routes between England, Africa, India and Australia. Alternatively, these travellers (assumed or implied to be British, or the Empire's White residents) were sufficiently senior in their diplomatic, military, scientific or commercial work to have their tickets paid for. Sir Alan Cobham's list comprised 'kings, princes, millionaires, sportsmen, business people, tourists of all ages and either sex'.[2] If England's upper class was the prime passenger market, at less than one per cent of the population it could only have generated a maximum 40,000 'PAX'. Approximately twenty times as many people (800,000) formed the English middle class in 1931, most of them earning between £250 and £500 annually.[3] The expatriate British air passenger market in India in 1931 was 110,000 men (the majority were commanding officers) and 45,000 women.[4]

Passage-paid or not, British emigrants (one third of a million) in the 1930s were certainly not air travellers. And the high fliers were definitely not only British. A 'miniature international assembly' was how Mr H. K. Hales (M.P. for Hanley, Stoke-on-Trent) described the passenger contingent on the first leg of his 1935 Empire flight. On board, for various sectors, were a photographer from London's *Times*, two Parsees returning to Bombay after their European tour, three Frenchmen, two

Italians, a Greek, two Indians *en route* to Delhi after giving evidence to a Select Committee, and two elderly American 'ladies' travelling 'unattended' to Geneva.[5] Reaching the British Empire from London, and *vice versa*, required European stops, and not all colonial residents were British. Resenting the considerable use Belgians made of the British airway to reach the Congo, an assistant in the Nairobi colonial secretariat once suggested that priority should be given to nationals of countries which subsidised the airway, and that other passengers should be charged higher fares.[6]

Passenger fares on *Imperial*'s single-class services clearly played a role in determining who flew, and how flying served and projected Empire. Rebates (of various kinds at different times) were available to some serving RAF officers and to some airline employees, for advance bookings made directly with the airline. School boarders qualified for discounts in 1938. Fares did drop in the 1930s, both absolutely and relative to rising real incomes.

Whatever the details, Empire flying was expensive. A flight from one end of the Empire to the other could cost half the annual salary of a middle-class Briton. *Imperial* charged £130 for a single journey to Karachi in 1929 and £98 (return £176) in 1931. This was only £6 more than the first-class surface fare for a journey that saved twelve days. Flying between London and Kisumu saved twenty-three days and was £2 less than the first-class overland fare. The ratios explain why Dr H. H. Hunter, the senior unofficial member of Uganda's legislative assembly, flew from Uganda to England in March 1931. At the time he represented the protectorate's commercial interests at a joint parliamentary committee on closer union in East Africa that sat in London. The African delegate to London did not travel by air. Three decades before becoming Kenyan President, the impecunious Jomo Kenyatta travelled by sea *en route* to promoting the views of his Kikuyu constituency.[7]

The steamship fare-class against which comparisons were made is crucial, and helps to explain the social class of air travellers. At the end of 1934 the single fare for the seven-day, 12,800 mile flight to Brisbane was £195, more than twice as expensive as the cheapest steamship fare of £96 and fifty per cent more costly than the £123 needed for a first-class berth. Similar ratios applied on the Cape Town run: £130 for a single air fare for the four-day, 7,900 mile flight, compared to a sea fare which ranged between £66 and £90.[8] The differential may explain the choices of a young cadet administrative officer in Africa. When going on six months' home leave in August 1934, Glyn Jones (later, Sir) used up his six-week travel allowance on a return overland rail journey between Northern Rhodesia and Cape Town, and a sea voyage to and from England. His journey out to what was his first colonial posting had been by sea, so his 'choice' was not about novelty.[9]

Urgency could be outweighed by cost in other ways. The last-minute decision by Gandhi to attend the Round Table Conference in 1931 meant that the ascetic Indian leader and his party of five could avoid paying an airfare that would have been unpalatably higher than the second-class return sea passage which they booked between Bombay and London.[10] On this occasion, the witticism about the expense of keeping Gandhi in poverty could have applied in as much as one airfare might have cost the least.

Some who flew as paying passengers along the Empire air routes did so in chartered light aircraft. In 1929, for example, the Vice-President of the Royal United Baking Powder Company of New York hired a small aircraft and pilot (Flt/Lt R. Bentley) to get him quickly from Johannesburg to Berlin.[11] Scheduled airline service had not yet begun on African sections of the route. Even after the start of commercial services there was some chartering along Empire routes. Toward the end of 1933 two doctors and a nurse chartered a light aircraft for a flight from London to Ahmedabad (going onward by rail to Bombay) and back to give urgent medical attention to the Prince of Nepal's wife.[12] When international air transport regulation was in its infancy, transit may not have been as easy as expected. As employees of an oil company discovered during their two-month, 19,000 mile marketing and business trip in one of the company's six- to eight-seat aircraft in 1935, frustrating and time-wasting delays due to customs and health formalities at airports between Cairo and Singapore were an incentive to use organised air services.[13] *Imperial Airways* occasionally chartered out its own aircraft. A four-month lease by Rhodesia and Nyasaland Airways accounted for three-quarters of £1,899 charter income in the twelve months to March 1936. The balance came from charters by the seventy-nine-year-old Chairman of the Prudential Assurance Company, Sir Edgar Horne (Cairo–Maan (Punjab)–Cairo), by a Senior Government official (Cairo–Maan–Baghdad–Damascus–Cairo) and the Palestine Government (Cairo–Ramleh).[14]

Official passenger records

Imperial Airways kept statistical records fastidiously. If they do survive in systematic lists, the personal biographical details of Empire airway passengers have yet to be found. Numerical data about passenger traffic are to hand, but are imperfect. In line with its opaque and secretive reporting, the airline used several styles of aggregate traffic reporting and showed its figures inconsistently. In corporate statistical summaries, data for 'Empire services' were often amalgamated with those for the (busier) European services. The airline also habitually expressed passenger loads in terms of passenger-miles, thereby concealing the actual numbers carried. There was no international aviation agency to insist on a standard rubric for collating and tabulating statistics.

A memorable working estimate is that 50,000 people used Britain's civil airline to get about the Empire in the 1930s.[15] Pointers to who flew on *Imperial Airways* are more scattered and reclusive than the partial record of passenger numbers. The airline's internal financial audits offer unusual glimpses (and confirm the social class of some of its clients). It was a Dr Weyersburg who lost a camera and £30 cash *en route* from Europe to Johannesburg and who was paid £1 11s 6d compensation.[16] The wife of the Under-Secretary of State for Colonies, Lady de la Warr (flying from Entebbe to London) and Sir H. A. F. Lindsay (flying from Kisumu to London) were among passengers who were sold incorrectly priced tickets in 1937.[17]

Perhaps the finest systematically collected long-run *Imperial* passenger record is in obscure British monthly intelligence reports from the Persian Gulf, but no Empire-wide view emerges from these regionally specific notices. They do at least share the authentic aura of official numerical tabulations, but inconsistency and overlooked detail are a limitation. Albeit incomplete, other qualitative information about inter-war Empire air travellers is rich and suggestive. Listings published (erratically) in some daily and weekly publications provide fragmentary clues to passenger identities. The lists are reminiscent of those printed so that socialite residents of port cities might know who had just arrived by sea. The Empire commentator Jan Morris remarked that the passenger lists which *Imperial* fed diligently to British newspaper columnists were not always very exciting. Two entries in a September 1935 list were at least exotic: the owner of the *Karachi Daily Gazette* and the Rumanian Air Minister.[18] Air passenger listings in the Singapore papers in the mid 1930s show that a high proportion of *Imperial*'s passengers to or from the city were only flying regionally.

Imperial's passenger listings appeared in February and March 1935 in the pro-Empire London weekly, the *Saturday Review*. They did not include all passengers. None were named off flights from India, or off connecting flights on the eastern route. The identification of passengers who had arrived in London via just the African airmail flight may have been incomplete: no passengers seem to have boarded south of Kenya. The passenger names that were printed indicate a predominance of White men. The names and titles (but not occupations) of at least some (booked) passengers on *Imperial*'s Africa service are buried deeply in the 'Personalia' columns of the weekly London paper, *East Africa*, from 1931.

Flight incident reports show men outnumbering women on *Imperial*'s long-haul flights, but the ratios were not static. A cinema newsreel shot at Croydon tells of five women among seven passengers on a February 1927 aircraft delivery flight for the Cairo–Karachi service.[19] British women passengers were generally given only vague identity if they were unmarried. Foreign men were scrutinised more carefully, being labelled, for example, 'American clergyman'. The title 'Italian fascist administrator' was printed proudly in one passenger's passport.[20]

Imperial losses

Aircraft accident reports are a useful source of information about Empire air travellers. No roll of the *Imperial* loss like the compressed one that follows was printed at the time. As the cold Board Minutes of *Imperial Airways* display, human tragedies were far away, far apart and beyond business.

All four passengers died when an *Imperial* flying boat *en route* to London sank in a gale on rough Mediterranean seas off the Tuscan coast in 1929. The *City of Rome* victims were not uniformly 'high class': a nurse returning from visiting her missionary family in India; a conservationist in the Punjab forestry department going on home leave to visit his wife and children; a customs officer at Croydon returning from holiday in Athens, and an airline employee. All three crew died.[21] Another *Imperial Airways* staff member – a ground engineer on transfer – was one of two passengers who lost his life in the *City of Jerusalem*'s fiery crash at Jask in the same year.[22]

Not all accidents involved fatalities. Sir Basil Blackett, a Director of the Bank of England and former finance member of the Indian Government, and Sir Hubert Young, Governor of Nyasaland, were on board *Artemis*, which was forced to land in jungle some eighty-seven miles from Mpika, Northern Rhodesia, in 1933. Haze forced the plane down. Having landed on marshy ground, it could not take off with the passenger load; two light aircraft were sent to collect the passengers and transfer them to Mpika, where they awaited *Artemis*.[23] Lord Balfour of Burleigh, travelling on business, was unhurt when the *Imperial* aircraft on which he and eight others were passengers hit a tree when taking off from Kisumu in November 1935 and was wrecked. There were seven men and one unmarried woman (flying from London to Brisbane) aboard the sixteen-passenger *Horsa* which put down on the Basra–Bahrain sector and had to be searched for. Among the seven, three were in the Indian Army and one was attached to the Indian

Medical Service. A thirty-two-year-old married father captained the four-man crew.[24]

No harm came to Mr G. Trace, *Imperial's* manager in South Africa, or three other passengers aboard the *Cambria* flying boat which landed in Kroosi Bay on the Mozambique coast after running out of fuel.[25] Shortly thereafter in 1939, and in the same vicinity, the *Challenger* flying boat crashed in Masuril Bay, Lumbo, after two aborted approaches. The aircraft sank in four feet of water but without passenger fatalities. One of the three passengers was Lt Col Kisch, former political officer for the Jewish agency. He was travelling from Alexandria to Durban. The other two passengers were a man (flying from London to Durban) and a forty-five-year-old advertising executive in the USA (flying from London to Lourenco Marques). The flight clerk (thirty-five) and the wireless officer died.[26]

An electrical engineer, the wife of a Cawnpore cotton merchant and an employee of the Anglo-Persian Oil Company were among those who died when the *City of Khartoum* crashed into the sea near Alexandria on the last day of 1935 after running out of fuel. The dead passengers included seven Britons, an American and an Italian. One, the employee of the Anglo-Persian Oil Company, was returning from home leave to the job he had held in Baghdad for twenty years. Another man was an engineer going to work in Australia. The cotton merchant's wife had been home to see her four children. She was one of two women aboard. The four crew who died were the pilot (thirty-three), flight engineer, wireless operator and cabin steward. Francis Yeats-Brown, more of a social 'somebody', had a lucky escape. The fervent imperialist and celebrated interpreter of India (he wrote 'Bengal Lancer') was going on a visit to the subcontinent he loved and would have been on the same flight had he not been turned back at Croydon.[27]

An air crash at Mirabella Bay, Crete, claimed the lives of two *Imperial* passengers in August 1936. One was a retired sixty-five-year-old former police officer in India who was returning to England. The other was a lieutenant (twenty-seven) in a British Army light tank division in India who was flying home to London on leave. The other five (male) passengers, and the crew of four, survived. A military officer serving in Egypt, an American clergyman and two other male passengers were injured. A banker, an aircraft designer and three high-ranking Indian Army Officers were passengers on the *Imperial* airliner that overshot Bahrain and flopped into the sand dunes within a week of the Mirabella disaster.[28] The sole passenger aboard the crashed *Capricornus* was a Folkestone clergyman's unmarried daughter who was going to visit her brother in South Africa. One of the flight deck crew who lost his life on the maiden flight was the son of the Director of Education in

Western Australia. Aged twenty-nine, he died on the first anniversary of his wedding.[29]

The three male passengers who drowned in the *Courtier* accident at Phaleron Bay near Athens in October 1937 were the commander of a bomber squadron in Iraq going home to England on leave, an American and a Greek. There was one married woman (flying from Basra to London) among the nine passengers, and three RAF personnel. None of the five crew was hurt. Injuries were sustained by a man flying from Basra to London, a squadron leader flying from Alexandria to London, and a Greek man flying from Alexandria to Athens. An RAF chaplain, Sqdr-Ldr Rev. Rees, and another male passenger were unhurt.[30]

A Government-nominated Director of *Imperial*, Air Chief-Marshal Sir John Salmond, was one of eight passengers (including one Indian and two women) who survived the *Cygnus* take-off crash at Brindisi in December 1937. Salmond had gone to India in the flying boat *Cordelia*, which was surveying the new flying boat route to Singapore. On his way back to England from Singapore, he had joined at Karachi. On the same aircraft, Robert Lutyens, the son of architect Sir Edwin Lutyens, was shaken but uninjured. An Australian passenger, a military captain who embarked at Charleville, and the twenty-seven-year-old *Imperial* steward were both killed.[31] The Royal Humane Society awarded Flying Officer Ralph Mountain its gold medal for rescuing three passengers.

Elites

Like crash victims, British royalty is conspicuously absent from lists in the glossy *Imperial Airways Gazette* and in the *Imperial Airways Monthly Bulletin*, two in-house publications directed at market making and news reporting. Foreign royalty were among the most celebrated passengers in the publicity *Imperial* gave to celebrity passengers. Few British royals flew to and from Empire countries in public aircraft. The practice must have irked *Imperial* for, as a marketing device, one royal name was attractive bait for another, and presumably the public at large would feel reassured by the readiness of royalty to leave the ground. Creating an image of class and style must have been a close consideration.

The significance attached to flying as a way of mobilising the Empire's great and good was expressed forcibly in 1937 by William Courtenay, the aviation writer and impresario. Wielding a purple pen, he suggested that the 'great Imperial crisis' that arose over the abdication of King Edward VIII in 1936 might have had a different outcome had all the Empire premiers who were summoned to London been able to arrive within a week. Even more extreme, Courtenay foresaw a time when British royalty would use aircraft to engage with the Empire. Within a decade, he suggested, the King would be able to fly to his most distant possessions sufficiently often for his subjects to see him many times during his reign. It was possible even to speculate that in future each dominion would hold a separate coronation ceremony, and that King George VI might personally open each dominion Parliament. First, he would have to be persuaded that his late father was wrong in his judgement that aeroplanes were nothing more than 'noisy, smelly contrivances'.

The idea of regal globetrotting had urgency about it. The Crown, Courtenay wrote, had become the last link of an unravelling Empire: the monarch was the magnet, a living emblem of 1,000 years of majesty and splendour. The sovereign was the mystic, spiritual link of Empire, a reminder of ceaseless royal power and benevolence. Air travel would be handmaiden to the new order, the agent of renewed life and power. Flight would prolong tradition, and the sovereign's air travel would strengthen and preserve the invisible bonds of Empire. The expectation that an English king might soon go aloft on Empire duty was anchored partly in the knowledge that before and after his coronation as King Edward VIII, the Prince of Wales had been a keen flier. The continuity delighted the air correspondent of the *Saturday Review*. He pronounced that the King's refusal to curtail his aviation interests foreshadowed the time when his kingdom would be bound to him by the greatest and busiest system of air routes in the world. Aviation would complement the merchant marine in making the Empire a prosperous and contented family. There seemed no reason that an airman king, a man whose aviation enthusiasm might be vital to the future prosperity and security of Empire, should not succeed a sailor king.[32]

For the associated status, it was fortunate for *Imperial Airways* that royals from other countries did fly in its aircraft. In 1927 the airline boasted having flown King Feisal of Iraq from Baghdad to Gaza. On another service the Emir Ghazi of Iraq flew from Cairo to Baghdad.[33] In 1937, the Aga Khan, holidaying with his wife in Nice, used *Imperial* to get quickly from Marseilles to Bombay, where his eighty-year-old mother, Lady Ali Khan, had suddenly become ill.[34]

In the early 1930s, *Imperial's* publicity highlighted several other royal trips. King Feisal chartered an airliner to fly him and his retinue from Baghdad to Cairo. It may have been a flight in more than one sense; the King died shortly thereafter in mysterious circumstances in Switzerland. A curious royal encounter made news as far away as Pretoria: King Feisal and his *aide-de-camp* played a hand of bridge with two Americans travelling by *Imperial* from India to England. The two Iraqis' identities only became apparent when they alighted at Basra, were met by a guard of honour and a band, and were driven away in a Rolls Royce.[35]

The Belgian Crown Prince, Leopold, used *Imperial* to help him reach and leave the Belgian Congo in the early 1930s. At least once he and the future Queen travelled together. Ex-King Ferdinand of Bulgaria flew from Aswan to Juba and, after three weeks of bird watching, flew back to Cairo. Leopold and Astrid were passengers on an *Imperial* flight from Entebbe to Cairo. At all landing stops the Belgian flag was flown from the aircraft cockpit alongside the Union Jack. Travel albums were getting stuffed with a new kind of photograph showing crew and royalty in front of a British civilian airliner at foreign airports.[36] John Spafford, the *Imperial* pilot who flew Feisal, Ferdinand, and Leopold and Astrid, was much taken by his encounters. He was also impressed by royal gifts and mementoes and by dinner in Ferdinand's private suite at Shepheard's Hotel. This was the high life indeed for the thirty-year-old Yorkshireman who left school at eleven, one of nine children from a humble working-class terrace in Scunthorpe.[37]

According to one report, the attention paid to King Albert I, Leopold's father, was irksome when he flew to the Congo in 1932. The *Imperial Airways* booking clerk was allegedly told not to make a fuss. In practice, 'there were special arrangements from one end of Mother Nile to the other. Wherever the big British bird alighted for a few minutes to leave a passenger or pick up mail, in popped a Briton to felicitate and annoy King Albert.' At Wadi Halfa a small launch plucked him off the airliner ahead of all other passengers, for whom he then waited on the shore, and to whom he apologised. Later in the flight, three RAF aircraft escorted the airliner to Butiaba. 'There, with a British band blaring "La Brabanconne" and "God Save the King" a right hearty British welcome was extended by His Excellency the Governor of Uganda, Sir William Gowers.'[38]

Persian Gulf ruling families were among *Imperial*'s passengers. The Emir of Tallal and his *aide-de-camp* were welcomed at Heliopolis after a regional flight in 1932. Similarly, the Sultan of Muscat flew from Gwadar to Karachi in mid 1935.[39] In 1934 Prince Ali Khan (son of the Aga Khan) made his second flight from India to Britain by air. The following year the Maharajah of Boroda and his party flew to London for the Jubilee celebrations.[40]

Other high-flyers in the inter-war period included the Egyptian Prime Minister who was photographed boarding an *Imperial Airways* plane at Cairo's Heliopolis aerodrome in 1932.[41] Gen. J. Smuts, then leader of the opposition in the South African Parliament, and two South African Cabinet ministers, flew by *Imperial Airways* from Cape Town to attend the world monetary and economic conference in London in 1933. During the overnight stop in Nairobi, Smuts addressed the specially postponed Union Day dinner of the Kenya South Africa Society. Despite early starts and heat, he wrote of his flight that it was 'an enormous success', punctual, restful and interesting.[42] A *Pathe Gazette* cinema newsreel showed him disembarking at Croydon after another flight in October 1934. Smuts returned to South Africa on *Imperial Airways* after a six-week speaking tour of Britain when he was also installed as Chancellor of St Andrews University.[43]

The Indian air route had its share of notable passengers. The Vicereine flew back to India on *Imperial* after four months' home leave in 1933. She did so again in 1934, with the Viceroy. In 1934 Sir Samuel Hoare met the homecoming couple (and an Indian cricketer who flew with them) at Croydon. Lord and Lady Willingdon's landing was captured by a *British Movietone* cameraman and inserted in the week's cinema newsreel. The *Aeroplane* and the smart London pictorial weekly, the *Sphere*, printed a photograph of the pair boarding an *Imperial* plane at Croydon on different occasions.[44]

Sailing to Britain on home leave in 1931, and returning permanently to Britain in 1933 by sea, Sykes would not have experienced what the Willingdons did while flying (with two male *aides-de-camp*) through the Persian Gulf in May 1934. Their passage to England created a new, if short-lived, ceremonial space on the ground. Dockside gave way to airfield site for public displays of honour. Flying had not ended Empire transport's role as custodian of pomp and ceremony which had previously been fulfilled by the P&O steamship line.[45] At Kuwait, a British naval vessel was specially commissioned to fire an extravagant fifty-one-gun salute when *Helena* touched down and a thirty-one-gun salute when taking off; the sea-fleet still had its uses. The Political Agent at Gwadar ordered the erection of tents on the landing ground for *Imperial*'s two-hour refuelling stop. A handpicked group of twelve Aga Khan boy-scouts kept guard at the canvas. The ground itself was 'ostentatiously picketed' by the Sultan's Arab askaris.

The Willingdons' welcome at Bahrain involved another thirty-one-gun salute and a dance performance. In the half-hour interval the Willingdons were led to the 'tastefully decorated' aerodrome shed where they met the Political Agent, the ruler of Kuwait, the naval commander, leading British and American residents, and Arab merchants and notables. Before departing, the Viceroy received a golden sword of honour from the ruling Sheik. Members of the public who happened to be on the flight would have basked in the reflected glow. Thirty-one guns boomed again at Bahrain in August 1934 when the Willingdons stopped on their return to India. On this occasion, after coffee and sherbet, the Viceroy reciprocated by presenting an ornamental clock to the Sheik. 'Ordinary Arabs' thronged the aerodrome at Kuwait to welcome one of their own leaders off an *Imperial* airliner in 1935. Three hundred police were no match for the 25,000 people gathered to

Figure 5 Dapper Sir Philip Sassoon, Under-Secretary of State for Air, boarding at Croydon for Africa, 28 September 1936.

greet the Sheik who, it was rumoured, had been enticed to London to be deposed.[46]

Not all prominent imperial personages flew with such pageantry. The lesser roll call included Mr H. U. Moffat, the Southern Rhodesian Premier, who used *Imperial Airways* on the non-Atlantic leg of his trip to the Ottawa Conference in mid 1932. Five years later his successor, Mr G. M. Huggins, flew by *Imperial* to attend the Coronation.[47] The head of the Egyptian delegation to the Coronation would also have flown had his direct request to the Egyptian High Commissioner succeeded: he asked Sir Miles Lampson's help to obtain seats on an *Imperial* flight. Lampson may have exhausted his influence with the airline by then. On the eve of the Anglo-Egyptian Treaty negotiations the previous year, 1936, he leaned on his London contacts to 'put in a word' with *Imperial's* headquarters 'to facilitate matters'. He was not asking to reinstate landplane service to Cairo's Heliopolis airfield so that he could avoid the 'great nuisance' of travelling to Alexandria (his daughter's flight from London landed at the new flying boat base there in 1935). He merely wanted seats on a flight to England at short notice during the busy summer season. Booking through normal channels had failed. Miraculously, seats became vacant. Lampson made no comment, as if he expected that outcome. His party of five that flew to and from London at either end of June included his wife, the Air Vice-Marshal and the Oriental Secretary at the Cairo Residency.[48]

Only a matter of two months before taking up his post as Governor-General of the Sudan, Sir Stewart Symes flew home to England from Tanganyika in November 1933.[49] With seven others (only half the passenger design capacity), Lady Somers flew aboard the inaugural commercial flight of the *Centaurus* flying boat to Australia in December 1936.[50] She went to join her husband, Lord A. H. T. Somers, England's Deputy-Chief Scout and President of the Marylebone Cricket Club, who was accompanying the English touring team at the time. It was a nostalgic return visit for the couple to a continent where he had been Governor of Victoria and Acting Governor-General five years previously.

Notables and officials

Rulers and statesmen were not the only precursors of today's jet set. Some people flew on British parliamentary business, or simply as a British parliamentarian. In 1929, Mr J. M. Kenworthy flew to India to advise Indian princes about the forthcoming Round Table meetings and the proposed campaigns of defiance and civil disobedience. Six male dignitaries in the British delegation from the Empire Parliamentary Association visited Tanganyika and Uganda in 1934 as *Imperial* guests. In January of that year Mary Rathbone, the fifty-nine-year-old feminist parliamentarian, flew from Brindisi to Gaza (in the company of her older half-sister) with the approval of the Colonial Office to inquire into child marriage in Palestine. Her previous trip to India (in January 1932) had been by sea. Seizing the chance then to study the depredations of Indian women's lives, and to apply corrective steps in the outcome of the final session of the Round Table conference, Rathbone could have flown. Instead, she used the lengthy voyage out to instruct members of Britain's Franchise Committee in the intricacies of the political status of Indian women.[51] Fast was not always best.

In 1933, the lists of *Imperial Airways* passengers included Sir John Maffey (switching from Governor-General of the Sudan to Permanent Under-Secretary of State for Colonies), Sir Auckland Geddes (brother of the *Imperial* Chairman, Sir Eric) and Sir Herbert Stanley. Respectively, their flights were from Uganda to London, London to Broken Hill and London to Cape Town.[52] Sir Pierre and Lady van Ryneveld departed on *Imperial* from Croydon in August to return to South Africa;[53] the journey was much easier for South Africa's Chief of General Staff than it had been thirteen years earlier when he became the first person to fly from London to Cape Town.

Mr H. K. Hales opted for the fastest possible transport for his business trip during Parliament's summer recess in 1935. Like others before him, he felt compelled to attach a wider message to the tale of his *Imperial Airways* journey to Baghdad, Karachi, Calcutta, Rangoon and Singapore (and beyond, on KLM to Java and Sumatra). Air travel, he concluded, 'is not only necessary but imperative if we are to retain our place amongst the nations of the world'. After his Africa return flight in 1936, Mr A. Montague, the M. P. for Leicester, remarked on the great work done by *Imperial* in knitting together Empire.[54]

Whether on assignment, on leave or taking up new postings, senior members of the armed services, the civil service and the Church were Empire air travellers. Mr W. M. Hughes, Australia's ex-Prime Minister and a participant in war Cabinet and early imperial conferences, would have wished that more Empire administrators were aerially mobile. In his published paean to Empire, Hughes wrote that only improved communications would enable imperial governments to 'exercise more effectively their rights to mould and direct the policy upon which the well being and safety of the Empire depend'. Advocating holding imperial conferences alternately in England and the dominions, he added that the presence of senior home government officials would 'arouse the interests' of dominion inhabitants in Empire affairs and 'would help them to realise they are Empire citizens'. Better transport would also give British ministers an opportunity to see the Empire themselves: 'with few exceptions, those who rule the Empire know very little about those great lands overseas whose destinies they shape'.[55]

Air Chief-Marshal Salmond flew eastbound through the Gulf on an *Imperial* service in 1934. On several separate occasions during the 1930s *Imperial* carried the British Resident at Bushire (Lt-Col H. V. Biscoe), his Under-Secretary and the British Vice-Consul at Ahwaz on home leave to England. Lt-Col T. C. Fowle, the British Resident at Bushire, used *Imperial* to fly him, his wife and their five-year-old son and his nursemaid on trips to New Delhi and London.[56] If they travelled with the boy when he posed with the captain on the steps of a departing airliner while on his fifth London–Basra trip, mother and nanny stood aside or were kept away from being photographed. The curious caption 'Trade Follows the Passenger' deflected any possible reading that thoughtful memento and effective publicity should not admit women into the air frame.

One of the earliest of the Empire airway passengers was Viscount Gage who, in 1930, flew on the RAF's 'India Air Mail' between Cairo and Karachi, and along the Nile section of the proposed Africa route.[57]

A member of the Uganda Medical Service, Dr M. Holliday was the first government official to use the Kisumu–London service in 1931; he flew with a Mrs H. Stone, the first woman passenger on the service.[58] The Colonial Secretary, Cunliffe-Lister (later, Lord Swinton), flew on errands to Egypt and Palestine in 1933, and to the Sudan and Kenya in 1934. Lady Cunliffe-Lister flew to Kenya to be with her husband after he became seriously ill. Their leisurely four-week sea voyage back to England was 'in sharp contrast to the discomforts of air travel'.[59]

Early in 1936, during his mammoth Carnegie-sponsored African survey, Lord Hailey dashed from Central Africa to London by air to meet briefly with his sick wife and the project steering committee. In 1937 Harold Nicolson (writer, broadcaster and Foreign Office official turned parliamentarian) returned home by air from East Africa after working ten weeks for a commission on African education. Associating aeroplanes with crashes, his wife, Vita Sackville-West, berated him for not taking the P&O steamship. He had certainly foregone the scale and style of the ocean liners that plied the Empire routes. British airliners were not floating hotels. Yet, even without fancy-dress balls, classical concerts and gaming tables they were nonetheless part of the moving tableau of Empire.[60]

The financial adviser to the Bahrain Government and the Financial Secretary to the Indian Government, Sir Alan Parsons, both travelled patriotically by air in 1934, one on home leave. Another *Imperial* passenger to fly on home leave (1937) was his majesty's minister at Jedda, Sir R. W. Bullard.[61] Rev. E. C. Stuart flew to England from Kampala in connection with his appointment as bishop of Uganda. Sir Hubert Young (but not his household retinue) flew out to Nyasaland to take up official duties as Governor.[62] In those postings it is doubtful that the availability of air transport swayed the decision to relocate, but there were pleasing prospects of more easily negotiated home leave and more regular contact with Canterbury and Whitehall. It is unclear what route Rev. Harold Wilde was following when, appointed as chaplain on Tristan da Cunha, he flew *Imperial Airways* to Kisumu.[63]

The alleged personal benefits of air travel revolved around the notion that time in transit was time wasted, but not all travellers would have resented a long, lazy sea voyage. And whereas it was desirable to avoid having journey time cut into leave periods, some employers excluded travel time from leave awards. In the 1920s, for instance, Kenyan public service officers were granted six weeks' fully paid travel leave in addition to normal entitlements. Saving some or all those payments in the public and private sectors would add to savings made by shortening temporary arrangements for leave replacements. Permanent staffing arrangements could also be adjusted. Some colonial business houses and government departments were said to hire four people for every three positions so as to cover home leave allowances.[64]

In 1935 the Governor of Uganda calculated that approximately twelve per cent of the salaries of important colonial officials was 'thrown away' by payments made for three months spent travelling to and from Britain every two years. There was no doubting his claim that air travel would reduce leave periods and the expense of keeping surplus men on the employment roll to cover for their absence (Uganda had seventy-seven officers filling fifty-seven posts), but whether the savings would be outweighed by the expense of air fares, surplus luggage tariffs and flight insurance was questionable. In addition to fretting about being charged with murder if an employee was killed after being compelled to fly home on leave (a revealing view of the comparative safety of air and sea travel at the time), the Colonial Office objected that the fatigue of flying undermined the recuperative value of an ocean voyage, and that shipping companies would object if air travel was made compulsory.[65]

If it was in employers' interests to have staff shorten their leave by flying, employees might have resented the way flying effectively truncated their spell away from work. Among those who may have felt cheated were a Kenyan District Officer, a Lieutenant in the King's African Rifles, the Director of Uganda's geological survey and a Northern Rhodesia District Officer. All four men flew at least one leg of their home leave journey before the end of 1932 when a standard regulation was adopted in the East African colonies which added eight days to the leave period at either end of a flight.[66]

Exactly how far the organisational advantages of Empire flight could be exploited for all government workers was a matter of cost, aircraft capacity and personal choice. The well-known imperialist Sir Evelyn Wrench (founder Secretary of the Overseas Club and Patriotic League, and Governing Director of the *Spectator*) broadcast on the BBC's Empire Service in 1935 that the mobility of Empire health officers might improve, and that some British statesmen might visit parts of the 'outer Empire' during their annual vacations. Reverse flying visits home by officials stationed in the colonies was already occurring. Air travel did not become mandatory for his majesty's agents going on overseas errands. None of the six members of Lord Bledisloe's Rhodesia and Nyasaland Commission, for instance, flew to or from central Africa on their three-month study tour in 1938. As for blue-collar workers, it was a gross exaggeration for Wrench to claim that air transport would make labour more mobile. Long before jumbo-size airliners, his claim was astonishing. Flight, he remarked, would make it feasible to transfer unemployed people from one part of the Empire to another to meet seasonal demands for labour. Industrial workers may well have preferred two or three weeks' summer harvesting in the Canadian West than spending their wages at English seaside towns like Blackpool or Scarborough, but the *hoi polloi* were grounded for some time yet.[67]

Entrepreneurs and professionals

The conjunction of oil and Empire provides evidence of senior industrialists and businessmen taking to the skies. In 1934 two senior employees of the Kuwait Oil Company travelled by air from England. Air travel enabled the London representative of the Bahrain Petroleum Company to visit his principals without leaving his city desk for more than seven days in October that year. The next month, *Imperial Airways* carried the (British born) Assistant Vice-President of the American Express travel company to the Gulf region to solicit oil company business. An American in charge of the Standard Oil Company's survey operations in the Gulf was an *Imperial* passenger in 1935. A medical doctor hired by the Company flew from Paris to Bahrain on *Imperial* to take up his post in 1937. In the same year the airline carried the manager of the Bahrain Petroleum Company, and his wife, on home leave to England.[68]

Oil company officers on brief assignments also used *Imperial* services to flit around the Gulf quickly. Similarly, Britain's political representatives in the Gulf flew often on the carrier's Middle East and Indian route sectors in the 1930s. Their regional business-trips, and those by sheiks, would not have been detected for mention in the London press or even the *Imperial Airways Gazette*. Neither would the flights by an accountant from Baghdad (auditing for the Bahrain Government), an *Imperial* site surveyor, the Gulf manager of the Cable and Wireless Company, the superintending engineer of the Delhi public works department or a Church of Scotland minister.[69] The resident manager of Bahrain Petroleum Company, and his wife, and the manager of the California Arabian Standard Oil Company went to Bahrain aboard an *Imperial* aircraft in December 1935.[70]

Some oil barons on important errands spurned air travel. Not even the abrupt cancellation of the Anglo-Persian Oil Company's operating licence in Iran prompted Chairman Sir John Cadman to dash to Baghdad in March 1933. Instead, he and two associates took a leisurely four-week sea journey from England to Iran via Bombay. Two other senior Company men travelled overland. Perhaps the relaxation of their journey set them up well for the crucial negotiations ahead. Cadman was not averse to flying. Undeterred by airliner crashes in September and October 1929, and October 1930, which unnerved Persians and aroused adverse comment about *Imperial Airways'* standards, Cadman was reported arriving by air at Bushire in January 1931. Then, in 1935, he flew from England to Iraq to attend the opening of the oil pipeline to Haifa. One of Cadman's fellow *Imperial* passengers was the bacteriologist Sir Alexander Fleming.[71]

A Director of the Armstrong Siddeley aircraft and motor manufacturing company left London aboard an *Imperial* flight in January 1932 to help establish a branch in Johannesburg. Mr E. C. Morse, President of the Export Division of the Chrysler Corporation, flew in each direction between Southampton and Durban on a five-day *Imperial* flying boat journey for a two-month-long visit to South Africa in 1938.[72] At a stage when only four 'ordinary' New Zealanders were said to have made the twelve-day commercial flight from Australia to England, Sir William Hunt, the prominent New Zealand businessman, used *Imperial* on sections of his flight from Australia to England and back for a five-week visit in 1937. He repeated the return air journey again in 1939, weeks before he died aged seventy-two. Mr F. Maurice Clarke, General Manager of New Zealand's Union Airways, flew imperially from Sydney to Southampton in 1938/39.[73]

Professionals with other paid careers also used *Imperial Airways* to move about. Surgeons, scientists and scholars were among the first converts to commercial air transport. Dr Humphreys bought an *Imperial* return ticket to Uganda late in 1931 when embarking on his year-long expedition to collect plants and seeds. Having served in the Royal Flying Corps (predecessor of the RAF), Humphreys would not have been a stranger to flying. One of his fellow passengers might have been the newly appointed General Manager of a Tanganyikan Estates company who left Croydon in December.[74] In 1933 the esteemed sixty-eight-year-old American archaeologist James H. Breasted chartered an *Imperial* aircraft to whisk him, his physically delicate wife, his son and an assistant round his old stamping grounds in the Middle East. Young Breasted (who had chartered an *Imperial* airliner the previous year to visit and film Persepolis) recollected that aerial views of the Fertile Crescent had given his father 'the supreme experience of his scientific career'. Rockefeller research money had bought the professor and his wife 'the most thrilling journey of their entire lives'.[75]

'A glorious experience' was how the revered British archaeologist, Sir Arthur Evans, described his first flight, across the Channel to Paris in 1921 with the Instone airline. He took to the air several times again in connection with his excavations at Knossos, Crete. In his eighties, Evans made his penultimate flight, in 1934, to Athens. He left Greece for the last time in a manner identical to Daedalus, the mythical Cretan. Other archaeological luminaries were warned off air travel by the fate of his son, Icarus. Two scientists with formidable reputations for fieldwork under *Imperial*'s flight paths across Persia and Palestine never acquired Breasted's aerial overviews. Sir Aurel Stein stuck to overland travel. And despite an invitation from *Imperial* pilots at Gaza, Sir Flinders Petrie never flew to his field camp or to his retirement in Jerusalem. The eighty-year-old acquired a bad headache after his trial cross-Channel flight in 1933, and it was this rather than a tight research budget that made up his mind.[76]

Using the modern tool, the famous anthropologist Bernard Malinowski used *Imperial Airways* to attend international conferences in Cape Town and Johannesburg, and to visit southern African field-workers in the summer of 1934. The airmail made possible a fortnightly turn-round of letters with his wife in London during his four-month absence. He returned home on the same flight as Gen. Smuts, whom he had already met in Pretoria.[77] One of Malinowski's protégés, Audrey Richards, planned to use the African airway to return to London a short while earlier. It was she who spotted at an airfield the contrasting sight of an irate business magnate alighting from an *Imperial* aircraft an hour late, and a mail runner preparing to trot off into the bush with his newly arrived load.[78] Two of Malinowski's other senior students, Godfrey Wilson and Monica Hunter, might have flown to and from eastern and southern Africa. Flight even eased travel for a Cambridge University undergraduate who returned to England by air in 1933 after an overland car journey from Johannesburg to Dodoma.[79] Air travel would certainly have helped put the new Rhodes–Livingstone Institute (1937) on the map. Siegfried Nadel, another of 'Malinowski's mandarins', surely flew from Nigeria to the Sudan when he was sent there in 1938 as anthropologist to the Anglo-Egyptian Sudan Government. In the same year, two American geographical experts booked seats on *Imperial*'s service to Alexandria. Their objective was to start their survey of the cradle of civilisation from the Nile to the Tigris/Euphrates.[80]

In December 1934 and 1935 the Royal College of Surgeons flew two of its members to Calcutta to conduct fellowship examinations.[81] Two geologists prospecting for oil in Australia and New Guinea were the first passengers to make and get a through-booking on *Imperial*'s London–Australia service in 1935: the first through-flight was fully booked by passengers on intermediate sectors before it was known that it would terminate at Brisbane.[82] Flying was the only way a Belgian research geologist could have reached Kivu Province, Congo, in time to study the volcanic eruption in 1938; he flew from Marseilles to Port Bell, Uganda, via southern Europe and Khartoum.[83] In 1936/37, an American surgeon and research scientist, and his wife, Grace Crile, flew to and from Nairobi on a zoological expedition. It was an easier journey than ten years previously when they had travelled overland to their 'jungle laboratory'.[84] Brig.-Gen. Sir Osborne Mance flew from London to Nairobi in July 1936 on a ten-week commission to study transport coordination in British East Africa. He drafted his report during a leisurely sea-voyage home.[85] On a year-long sabbatical from Cambridge University, the medieval historian Helen Cam flew to and from India on *Imperial Airways* in 1936/37.[86]

On flying business

People connected with *Imperial Airways* and those conducting aviation negotiations or doing airline business also travelled about the Empire by air. The self-serving circularity of flying so as to promote and protect interests generated by flight itself may not have been anticipated by the proponents of imperial aviation, but the missions served Empire nonetheless. *Imperial* fared least well: transporting its senior staff free meant losing passenger revenue over and above that shed because some government officials were flown at reduced rates.

In 1929, after completing his *Imperial Airways* mission, Lord Chetwynd flew home from India with his daughter on the airline's first return air mail flight. The following year, before the *R101* airship crash and his appointment as Sir Sefton Brancker's successor, Sir Francis Shelmerdine returned to India by air after attending a conference in London. Later, as newly appointed Director-General of Civil Aviation, he returned to London with his wife by sea. The decision may have had to do with rest, or it may have been dictated by the outcome of ministerial quibbling about his salary, untaken leave, interrupted contract and air fare. Shelmerdine's successor in India, Tymms, flew to London and back in 1932 to discuss ways of pushing the Empire airway east across India.[87] Air Vice-Marshall A. W. Tedder's flight in July 1938 had another imperial purpose: he left his RAF Command in Singapore in an *Imperial* flying boat bound for England to take up the position of Director-General of Research Development at the Air Ministry.[88] On the longer Australian haul, Hudson Fysh, the Managing Director of Qantas, which organised the air route south of Darwin, flew to London in mid 1933 to negotiate a partnership with *Imperial*. On his first journey abroad since his wartime flying service, Fysh was enchanted and impressed by the readiness with which sterling currency ('the promise of England') was accepted along the route and by 'the fine Empire building work' carried out by *Imperial Airways*.[89] In the summer of 1935, Mr G. L. Gandy, of the Directorate of Civil Aviation, left London by *Imperial* for New Delhi to take up his appointment as Deputy Director-General of Civil Aviation in India.[90]

Flying commended itself not least to senior Air Ministry officials who on several occasions used *Imperial* in the course of last-ditch foreign negotiations. Personal meetings were resorted to after written correspondence failed. In 1935, Sir Christopher Bullock, an aide and three technical representatives from the Air Ministry, flew to Lusaka, Nairobi and Khartoum. Their three-week tour was to negotiate details of the Empire Air Mail Scheme (EAMS) over which the South Africans in particular were being obstructive. In his meetings Bullock pointed to the benefits of aviation by speaking about the prominent Khartoum businessman who happened to be on the same outbound flight having tendered for orders in Britain.[91]

Also in 1935, four senior officials from the Air Ministry and the Post Office flew to India and Australia to thrash out details of the EAMS as it affected those countries. Two years later, Shelmerdine flew to Rome in a three-day effort to wring concessions from the Italians regarding the use of Lake Bracciano by civil flying boats.[92] Political types flew, and so too did technical types. Toward the end of 1931 the person sent to survey wireless organisation on the Cairo–Cape Town route travelled on a civil aircraft delivery flight; *Imperial* allowed slight diversions to allow him to visit remote places.[93] The Assistant Director of Works in the Air Ministry flew to Durban in 1936 to advise on the construction of the flying boat base there.[94] In 1938, an Air Ministry official, Mr W. L. Lang, occupied a seat on *Imperial*'s flying boats during his eight-week inspection of the Empire air mail route through Africa.[95] Having flown quickly from London to Singapore by KLM, Lt-Cdr A. J. Tillard used *Imperial* wherever possible on his three-month survey of the Alexandria–Singapore flying boat route in 1938/39.[96]

Imperial's Operations Manager, Mr A. C. Campbell-Orde, made a 31,000 mile inspection tour of Empire air routes in mid 1939. His forty-six-day tour involved 208 hours' flying. Shortly thereafter, Maj. Newman, the airline's foreign public relations officer, travelled to the Near East. In the same season, *Imperial* carried a civil engineer out to Kisumu. The airline hired him to survey fluctuating water levels in Lake Victoria's Kavirondo Gulf and to assess relocation of the Kisumu flying boat base from there to Lake Naivasha.[97] Three engineers (plus food and supplies) were aboard *Cordelia* when it was sent from Singapore to Siam to assist its stranded sister flying boat *Cassiopeia* in July 1938.[98]

In a fast-evolving imperial order it may have been a sign of the social acceptance of flight that neither Bullock's nor Shelmerdine's mission attracted attention in Britain's daily press. Yet, even in the infancy of the organised Empire civil air service, the departures of leading public servants on important duties went unnoticed. No officials, it seems, saw any merit in leaking notice of flights that would draw attention to the difficulties of organising imperial aviation.

No pencils scribbled when Brancker and Burchall flew to Greece in 1929 to negotiate a right of way for *Imperial* at the start of the first commercial London–India flight. Burchall's trip to and from Nairobi in May 1931 claimed less attention than his presence and remarks there.[99] No cameras clicked when Woods Humphery flew from India to Cairo in 1930 with the Deputy Director of India's Post and Telegraphs Department to meet with British counterparts in an effort to overcome obstructions that threatened his airline's subsidy-linked overseas air mail mandate and revenues. Britain's Post Office Secretary left Croydon unnoticed.[100]

Only African newspapers reported flights made in 1939 in connection with *Imperial* taking over Rhodesia and Nyasaland Airways and Kenya's Wilson Airways. Cmdr B. W. Galpin, an assistant to *Imperial*'s General Manager, flew back to London from Durban after accompanying one of the airline directors, Sir John Maffey, on a tour to consult British governments in Africa and the South Africans.[101] The *Imperial* flight by the aircraft manufacturing company's five-man salvage team and its own ground engineers to a crash site in the Belgian Congo in 1939 was not publicised. Giving free return air tickets to Europe to the local Belgian Provincial Commissioner and his wife as a token of airline gratitude for assistance during the ten-month-long rescue meant foregoing yet more passenger revenue.[102]

Last in the category of Empire airway officials who were carried by *Imperial* were the airline's own operating staff. Their relocations were to do with initial and subsequent postings as *Imperial*'s route network grew, and with leave and leave replacements. Indicatively, nearly twenty-five per cent (659) of the 2,777 passengers on *Imperial*'s London–Karachi services between April 1930 and March 1932 were airline staff. Seventy per cent (1,938) were paying passengers, and a little more than five per cent (155) flew for free. On the Africa services, 380 of the 3,932 passengers in the accounting period April 1931 to March 1933 were airline staff, but the percentage was over twenty (213) in the 1931/32 financial year. Three-quarters of passengers (2,948) paid and eight per cent (315 passengers) flew without charge in the two-year accounting period.[103]

Employee movements were reported avidly in *Imperial*'s low-budget staff news sheet, mimeographed weekly. Listings show that in January 1934, for example, thirty-five airline personnel took up seats on the Empire routes. In May 1936 their number was twenty-two. Twenty-six employees flew in each of June 1936 and September 1937. In March 1937 thirty-three airline employees either displaced revenue-paying passengers or filled vacant seats. Wireless operators, aircraft ground engineers, coxswains and station superintendents appear to have been most mobile. Aircrew and ground engineers, as well as pilots and cabin staff, also flew on aircraft delivery and repositioning flights.[104] Storekeepers and accountants flew on occasion. *Imperial*'s medical adviser, Col F. P. Mackie, flew from London to Karachi in 1937. The carrier's chief catering officer did a tour of inspection on the India route in 1934. Maj. Hodge, *Imperial*'s Station Superintendent at Pietersburg (South Africa), made his last flight to Germiston *en route* to a Johannesburg hospital for an amputation. He died in flight.[105]

Two of *Imperial*'s engineers (one assigned to Sharjah, the other to Singapore) were the only other passengers on a flight which the writer Eileen Bigland made from Southampton to Rangoon and back by *Imperial* flying boat in 1939. Sharjah was not a favoured posting, and the engineer going there guessed that he'd not be long: 'the war will rescue me', he remarked.[106] By then, as air mail loads increased, the shortage of seats on *Imperial*'s scheduled flights had led to restrictions on paid passages of the family members of airline employees on overseas duty. From June 1938 *Imperial* paid only for the surface passage of the wife (and not more than three children younger than sixteen) of permanent overseas staff members and those returning on normal leave at the end of their first and subsequent contract or those on permanent transfer.[107]

Passenger diversity

Even if the caricature of the 'typical' *Imperial Airways* passenger included the airline's non-paying customers, it would be one that was dominantly male and middle class. Eileen Bigland met only four other women passengers (including 'a titled lady') on her London–Kisumu return flights in 1938. The fifteen passengers with whom she travelled back to England included two natural history experts who had been photographing game in Kenya.[108] The paying passengers were certainly a more diversified set than social elites and corporate leaders. There is some sense that the profile of passengers widened on the Empire services as the number and capacity of services grew, and as *Imperial*'s route network expanded. The occupational roll changed more than the socio-economic tier.

Hunters of big game were among the wealthy perambulators by air in the 1930s. Leaving London in May 1932, Mr N. M. Kater, a Sydney resident, flew to Africa on a hunting trip along *Imperial*'s route. He spent one month at Nairobi, two weeks at Mbeya and three weeks at Pietersburg (South Africa).[109] Joy Packer's account of her flight from London to Cape Town in 1934 mentioned a hunter bound for Kenya (also aboard were two nuns going to Juba and a young male cotton purchaser).[110] An *Imperial* manager in Egypt recalled that in the 1930s

Figure 6 *Imperial Airways* passengers and Shilluk men at Malakal (Sudan), 1936.

roughly two per cent of passengers were going on safari, mostly to Kenya and Uganda.[111] Lady Evelyn Cobbold's son (who flew to and from Kenya with her in 1934 on an extended trip) would have ended up in that category. The passengers they encountered included five women (including a renowned American monologue artist, a lady and her maid, and an octogenarian without her maid) Maj. Grogan, Mr and Mrs de Havilland and a retired naval man.[112] Age was no barrier to flying. Lady Cobbold was sixty-seven (and had undertaken a pilgrimage to Mecca in April 1933). In 1932, the seventy-year-old Dame Ethel Locke-King flew to visit her nephew in Northern Rhodesia. At seventy-five, Col S. Paterson returned to Kenya by air after visiting England.[113]

Among others who travelled in connection with wildlife was Miss Ruth Rodgers, who received air tickets from her mother as birthday gifts. Her flight from Croydon to Entebbe and Mbeya in 1935 was not her first: she had already flown 5,000 miles by *Imperial*, winning the Institute of Amateur Cinematographers gold medal in 1934 for a film taken during one of her flights. Her 1935 trip was to film wildlife, and also to photograph Africans and their customs.[114] Later the same year, Miss K. McElwee flew return from London to Kisumu for a month-long safari in Kenya and Uganda. Three years on, Mr Lincoln Ellsworth, an American explorer, also flew to Kisumu. After filming wildlife he travelled to Cape Town to begin a South Polar scientific expedition. Sqdr-Ldr T. C. Pattinson used *Imperial* from London to Kenya and back in 1939. After his six-week hunting safari he flew home with his 'bag' of elephant ears 'to use as table tops'.[115] Anglers were a slightly different category of hunter: Dr and Mrs Sutton, ardent fishers, flew from Durban to Southampton on their way back home to America.[116]

Spotting journalists and curious writer-travellers on an *Imperial Airways* flight was not unusual while there were still new routes being opened and when new aircraft were flying new schedules. But when even the most senior Empire newspapermen needed to travel in any numbers, the limited capacity of aircraft was a deterrent. Stella Wolfe Murray was the only journalist on the first *Imperial Airways* flight from Egypt to the East.[117] All forty-two delegates and their thirty-three companions travelled by sea to and from Cape Town in 1935 for the fifth Imperial Press Conference.[118] In a smaller group (which included two paying passengers), Muriel Howlett, a BBC newsroom secretary, was one of *Imperial*'s nine guests on the inaugural flying boat through-service to Australia in mid 1938. She was to record her impressions in radio broadcasts from Calcutta, Singapore and Sydney. The rapidity of her return to the country she left in 1935 must have been striking compared to her six-week-long northbound sea journey.[119] *The Times* journalist assigned to cover the Sino-Japanese conflict for six months in 1938 returned to London from Hong Kong via Singapore on *Imperial*.[120]

A playwright who spent ten days flying to London on six aeroplanes across ten countries in 1935 had already commented on the quickening of travel from Australia to Britain. Claiming to be the first Australian woman on a commercial flight to London, she wrote up her experiences in a series of articles in the women's supplement to the *Sydney Morning Herald*.[121] Other women followed. Miss Hilda Green, an Australian schoolteacher, took a return flight to England during a 1937 school vacation. Mrs M. Symonds, a middle-aged Sydney resident, flew to England before sailing to and from the United States for a two-week visit, then flew to Scotland to see friends before taking the *Imperial* flying boat home. Her 35,000 mile journey took less than eight weeks.[122] Flying across continental interiors, and making brief stops in isolated backwaters, could never have the same effect on young Australian women as had the more common 'secular pilgrimage' to England by sea. Self-identification tied to languid, sociable ocean voyages would be different. So too would be the notion of Empire which was acquired (and was projected) away from the bustling port cities on its watery rims.[123]

One of the journalists aboard the same aircraft as Howlett was William Courtenay. During a twenty-minute radio broadcast in Sydney he confirmed that there were no public celebrities aboard. One woman was mistaken as 'a university professor of geography in Copenhagen' (she may have been a professor's wife, using a form of address such as 'Professorinde' or 'Fru Professor').[124] On various legs of the flight there were several British Army officers returning from leave to India's northwest frontier, an Australian sheep farmer and his wife on a world tour, a ten-year-old schoolboy going on holiday, and 'sundry gentlemen of India, Burma and Siam'. The uneven identification of passengers suggests different conversational engagement at best, and deliberate detachment at worst. In the four partitioned flying boat cabins it is possible that Australia's Director of Posts and Telegraphs, its Chief Inspector of Postal Services, a Scotsman who had been in Java and a Sydney barrister could have avoided a father and his pathetic-looking son, a 'coal-black sammy' – a social 'nobody' if ever there was one.[125]

A social self-selection process based on high fares meant that *Imperial* rarely violated the class-consciousness and social stratification in England and Empire in the 1930s. The experience of a déclassé 'Jack Tar' on a 1938 flight must have been excruciating. It would have helped little that he was (or said he was) 'no ordinary hand' but a leading stoker. Sniffy passengers may have been more sympathetic knowing that his messmates had helped finance his flight on receipt of some (unspecified) 'grave news'.[126] Even had he sat away from other flying boat passengers, a well-scrubbed and smartly dressed ship's coaler would have struggled to conceal his bearing, manners, accent and conversation, the tell-tale signs of social class.

Some social mixing of passengers was evident on flights to and from Africa from the start. The Chairman of the *East Africa Standard* newspaper flew from Nairobi to London in November 1931; the Director of Carr's biscuit manufacturers flew the same route five months later.[127] Two of the eleven passengers on an *Imperial* flying boat from Juba to Khartoum in 1932 were a couple who had discovered that air tickets costs £10 less than Nile steamer tickets for the same journey. *En route* from Durban to Cairo by motor caravan, they consigned their vehicle by barge and travelled for twelve hours instead of eleven days. In January 1933 a Johannesburg lady and her daughter flew to England on *Imperial*, and 'an energetic northern Transvaal stock breeder' used the air mail service to get to and from England in less than twenty days.[128]

In 1936/37, the Criles' companions were a copper-mining engineer, a sea captain going to Durban, a Harrogate physician travelling to Cape Town and a woman from Berlin. On their return journey, the Criles travelled with a Scottish engineer who worked in Johannesburg.

Beyond Alexandria, they shared the *Imperial* plane with a British woman journalist, an army major returning from Quetta, a seventeen-year-old English boy who was returning to England from Delhi for the first time in seven years, and a young woman who worked at an Indian mission station. A similarly diverse group of twelve passengers was on a southbound *Imperial* flight into Entebbe, Uganda, in 1936. The seats were occupied by a coffee planter, businessmen bound for the Cape, government officials returning from home leave, an English peer going to his Kenyan estates, a female lepidopterist and members of a commission of inquiry.[129] Those travelling without clear purpose aboard this flight were tagged 'sightseers', a category which actually included all aircraft passengers.[130] The two Johannesburg residents who bought return air tickets on *Imperial* to spend one week in London in 1935 were identifiably tourists.[131]

An early instance of passenger diversity on the Empire routes involved the film crew which left London for Africa to shoot the colonially themed film 'Sanders of the River' in 1933.[132] Air travel helped make out-of-the-way places accessible, and also reduced the amount of time spent by employees on activities for which clients could not be charged. *Imperial Airways* was drawn into filming even more directly when it cooperated in the making of the film 'Air Liner' in 1935 and flew the associate producer to Cape Town.[133] In 1937 a Hollywood team flew to Kenya to make the film 'Stanley and Livingstone'. Staff from New York's Museum of Natural History flew that way in 1938 to photograph game.[134]

In the absence of official information, the presence of journalists aboard Empire flights was handy for history. When, in 1930, an American news reporter rushed to interview Mahatma Gandhi, the Indian political and spiritual leader, he filed a list of other people on one of the early *Imperial* flights to India. Occupational diversity exceeded diversity of gender and social class. In the company of the airline's chief accountant the reporter left Croydon with an American businessman *en route* to Vienna and a representative of a Glasgow shipbuilder bound for Athens to negotiate a contract. The titled lady of sixty-three on the flight was dashing to Basra to be with her critically ill niece, and the Deputy-Inspector of Police in northwest India was hurrying to Peshawar.[135]

If one flight record is indicative, the passenger profile remained mostly male and middle class into 1937. During an England–Australia return journey, a staff writer on assignment for *Flight* magazine sat with the manager of a teak plantation in Burma, an American engineer and a captain in the Seaforth Highlanders who was returning to Britain from Hong Kong. A prominent Australian importer in the confectionery business was also aboard. One or two of these passengers may have been sufficiently 'Poona' to dress for dinner in the air, a style that William Courtenay heard about during his flying boat journey to Calcutta and back in nine days in 1938.[136]

The results of a passenger survey (period and geography unspecified) that *Imperial* itself undertook in 1937 convey a similar pattern. Coinciding with some fraying of social cachet, a list publicised in a talk on BBC radio and in an article in *Chambers's Journal* identified several non-British nationals. The Australian passengers included two jockeys, a Sydney cotton spinner going to England to buy new machines, a circus manager arranging a tour in the East, and a radio engineer returning from a study of European television. The Americans on board were the export manager of a pen company inspecting Singapore agencies, a retired boot manufacturer (and his wife) on a world tour, and a honeymoon couple. Other passengers were a Dutch consulting engineer, a Chinese tailor from Darwin visiting relatives in the Malay States, engineers going to investigate oil and mineral possibilities in Malaya, two mining engineers returning from leave to Siam, and a New Zealand farmer going home.[137] An unusual category of passenger was the patient whom *Imperial* rushed 8,000 miles from Durban to London in six days for an urgent operation.[138]

In 1938, a British district engineer in the Sudan Public Works Department returned by flying boat (with just one other passenger) to his post after a period of home leave. In June of that year it was reported that forty per cent of the Empire passengers handled at Southampton were 'businessmen'. Thirty per cent were government officials (including those starting and finishing their home leave). Twenty per cent were passengers travelling for pleasure. Five per cent were making urgent private journeys. Five per cent were unaccounted.[139] By the late 1930s, indeed, *Imperial* seemed pleased that it did not just cater for elites and was performing a broader Empire service. In 1939 the airline proudly reported 'passengers from different walks of life': a musical conductor outbound to Australia, a woman going out to Africa to get married, a woman flying to the bedside of a sick relative, a hunter, and members of an expedition seeking rare plant and animal specimens in Burma.[140]

Such *Imperial Airways* passenger details as can be retrieved from discrepant sources speak to the geographical expanse of Empire and the diffuse business and administration associated with it. British men with young families were among the most scattered of the Empire's agents. Some were fortunate enough to stay connected by air with fiancés, wives and children. Miss Joanne Briscoe George (and her mother) took an October 1935 *Imperial Airways* flight from London to

Figure 7 Pamela Cross, her mother and *Imperial*'s flying officers disembarking at Galilee (Palestine), October 1931.

the Sudan, where she was getting married.[141] In April 1939 an officer in the Colonial Medical Service in Nyasaland met his future wife at Chileka airport after her six-day journey from Southampton to East Africa by flying boat and then local flight. The six-day journey left her 'worn and wan'.[142]

Young married women were precisely the kind of person *Imperial*'s medical adviser had in mind when seeking clarity from experts in 1937 on whether and how to regulate 'the transit of pregnant women by aeroplanes'.[143] In 1933 a ten-week-old baby arrived at Croydon from Basra with its mother to spend summer in England. Mrs B. T. Place and her infant son flew from London to Entebbe in 1936. After the wife (twenty-three) of a doctor in East Africa's government service flew to London in June 1938 to give birth, she returned six weeks later by air to Mombasa with her new son. *Imperial* was reported to have constructed a screened bunk for the infant, and to have picked a 'ship steward' specially to act as 'nurse' for the trip. The mother's view was that a month-long sea voyage through the Red Sea in summer would have been much more terrifying than air pockets.[144]

Little speaks better to social separation in the Empire (and to views about where education was best) than the children who flew. In 1933, aged eleven, P. Rydon (son of an M. P.) flew unaccompanied from Nairobi to London for school. David MacDonald was already a 'veteran' flyer when, aged five, he made his fifth flight with his mother from London to Baghdad to visit his father who worked for the RAF there.[145] John Fowle was five when he made his fourth trip to Kuwait with his parents. As long as his father remained the Political Agent there, he would – and did – fly more.[146] An eight-year-old was aboard an *Imperial* flight from Cairo to Khartoum in 1935.[147] Another young boy on his second annual England–Kenya summer holiday flight made the news in 1938. In 1939, a ten-year-old girl flew from Kampala to Southampton to attend school in Winchester; a recorded interview recalls a schoolgirl's flying boat journey to India in the same year.[148] In 1934, the son of Maj. Maxwell (a well-known Nairobi resident who made at least two return flights to London) flew to Kenya for Christmas. The following year, a teenager flew with two other passengers on an aircraft delivery flight: the fifteen-year-old son of Sir H. Stanley, the Governor of Southern Rhodesia, was going to spend summer with his parents.[149] Care of orphans was another matter: a Mr Richards was filmed arriving at Croydon with his 'motherless son'.[150]

Notes

1 J. Morley, *Colonial Postscript: Diary of a District Officer, 1935–56* (London, 1992), p. 3.

2 *Bystander* (25 May 1932), p. 396.

3 McKibbin, *Classes*.

4 M. A. Procida, *Married to the Empire: Gender, Politics and Imperialism in India, 1883–1947* (Manchester, 2002).

5 H. K. Hales, *The Chariots of the Air* (London, 1936), p. 15.

6 NA, CO 822/48/1: J. Barton to Colonial Office, 17 July 1933.

7 *The Times* (30 March 1929); K. Ingham, *The Making of Modern Uganda* (London, 1958). A fourteen-day time saving on the London–Kisumu routes was reported by E. Kettles-Roy, 'To East Africa by air', *Air and Airways*, 8 (1931), 297–8.

8 *Economist* (29 December 1934).

9 C. Baker, *Sir Glyn Jones: a ProConsul in Africa* (London, 2000).

10 *The Times* (31 August 1931).

11 *Cape Times* (12 July 1929), p. 11; *Star* (6 August 1929).

12 *Flight* (28 September 1933), p. 981; (2 November 1933), p. 1101.

13 *Flight* (19 December 1935).

14 British Airways Archive and Museum (Heathrow) (BAAM), AW/1/4040(2): *Imperial Airways* Profit and Loss Account, Financial Year Ended March 1936.

15 For detail and discussion see G. H. Pirie, 'Passenger traffic in the 1930s on British

imperial air routes: refinement and revision', *Journal of Transport History*, 25 (2004), 66–84.

16 BAAM, AW/1/4043(1): Accountant, Southern Africa Area, to *Imperial Airways* Secretary, 1 April 1938.

17 BAAM, AW/1/4043(1): Internal Auditor to Managing Director, 29 June 1937. Report no. JSS/16.

18 J. Morris, *Farewell the Trumpets: an Imperial Retreat* (London, 1978), p. 359.

19 *Pathe Gazette* (23 February 1927).

20 P. E. Sims, *Adventurous Empires: the Story of the Short Empire Flying Boats* (Shrewsbury, 2000); *Manchester Guardian* (2 January 1936), p. 9.

21 *The Times* (29 October 1929), p. 16; *Manchester Guardian* (28 October 1929), p. 9.

22 BL, OIOC, L/E/7/1565 (6579); *The Times* (9 September 1929).

23 *Manchester Guardian* (14 June 1933), p. 9; *Daily Mirror* (14 June 1933).

24 *Manchester Guardian* (4 November 1935), p. 13; (31 August 1936), p. 9.

25 *Cape Times* (9 March 1939), p. 11; Sims, *Adventurous*, pp. 114–18.

26 *Manchester Guardian* (2 May 1939), p. 11; Sims, *Adventurous*, pp. 132–5.

27 *Manchester Guardian* (2 January 1936), p. 9; J. E. Wrench, *Francis Yeats-Brown* (London, 1948).

28 *Observer* (23 August 1936); *Manchester Guardian* (24 August 1936); (31 August 1936), p. 9. Sims, *Adventurous*, p. 62.

29 Sims, *Adventurous*, p. 59; *Evening Post* (Wellington) (27 March 1937), p. 13. Gold bullion rumoured to be aboard *Capricornus* may have been destined for Indian royalty.

30 BAAM: *Imperial Airways* Board Papers, N 27 (246); *Manchester Guardian* (2 October 1937).

31 *The Times* (6 December 1937); *Manchester Guardian* (6 December 1937), p. 9; Sims, *Adventurous*, pp. 67–8.

32 Courtenay, *Airman*; *Saturday Review* (1 February 1936), p. 133; Williams, *Something*.

33 K. Hudson and J. Pettifer, *Diamonds in the Sky: a Social History of Air Travel* (London, 1979); *Imperial Airways Bulletin* (September 1927).

34 *Observer* (21 November 1937), p. 20.

35 *Imperial Airways Gazette* (October 1932; July 1933); *Aeroplane* (December 1932).

36 *Imperial Airways Gazette* (March 1930; April 1932; May 1933); *Airways and Airports* (August 1933), p. 156.

37 J. Spafford, *The Comings and Goings of John Spafford* (privately published, 1989).

38 *Time* (11 April 1932).

39 *Popular Flying* (November 1932), p. 470; *Imperial Airways Gazette* (July 1935).

40 *Imperial Airways Staff News* (7 February 1934); *Imperial Airways Gazette* (June 1935).

41 *Imperial Airways Gazette* (March 1932); *South African Motorist* (November 1932).

42 *The Times* (24 May 1933); Manuscript and Archives Department (University of Cape Town), Duncan Papers, BC294 (A5.15.2): Smuts to Sir Patrick Duncan, 15 June 1933; J. van der Poel (ed.), *Selections from the Smuts Papers, Vol. 5* (Cambridge, 1973), pp. 563–5.

43 *Aeroplane* (7 June, 12 July 1933); BNIS, *Pathe Gazette* (8 October 1934); *African World* (November 1934).

44 *Bystander* (1 May 1934); *Sphere* (18 August 1934). The cricketer Duleepsinjhi had previously flown to India to attend his brother's wedding: *Flight* (20 June 1934), p. 1660.

45 Morley, *Colonial*.

46 PDPG (May 1934; September 1935). The British Empire and Commonwealth Museum holds brief film footage of the Willingdons' arrival on board an *Imperial Airways* aircraft at Al Muharraq, Bahrain on 17 May 1934. See media clip #6628 at www.imagesofempire.com (accessed August 2010).

47 *Flight* (17 June 1932), p. 523; *Saturday Review* (8 May 1937), p. 314.

48 *Imperial Airways Gazette* (August 1936), p. 7; M. E. Yapp (ed.), *Politics and Diplomacy in Egypt: the Diaries of Sir Miles Lampson, 1935–37* (Oxford, 1997), pp. 579, 802.
49 *The Times* (18 January 1934), p. 13.
50 *Argus* (Melbourne) (18 January 1937), p. 9.
51 J. M. Kenworthy, *Sailors, Statesmen-and Others* (London, 1933), p. 239; *The Times* (15 November 1934); M. D. Stocks, *Eleanor Rathbone* (London, 1949).
52 *Flight* (30 November 1933).
53 *Flight* (7 September 1933), p. 902.
54 Hales, *Chariots*, p. 164; *Imperial Airways Gazette* (August 1936).
55 W. M. Hughes, *The Splendid Adventure* (London, 1929), pp. 270, 280.
56 PDPG (November 1929; May, December 1931; August 1932; March, May, July, October, November, December 1934; March 1937).
57 *Hansard* (Lords) (3 June 1930), col. 1342.
58 *The Times* (15 May 1931), p. 16.
59 J. A. Cross, *Lord Swinton* (Oxford, 1982), p. 131.
60 J. W. Cell, *Hailey* (Cambridge, 1992), pp. 224–7; N. Nicolson (ed.), *Harold Nicolson: Diaries and Letters, 1930–1939* (London, 1966), pp. 18, 295; M. Drabble, *Angus Wilson* (London, 1995).
61 PDPG (April, October 1934; July 1937).
62 *Air and Airways* (December 1934), p. 328; *South African Motorist* (January 1933).
63 *Imperial Airways Weekly Air News Bulletin* (14 December 1937).
64 Hudson and Pettifer, *Diamonds*; *Air and Airways*, 8 (1931), p. 298; *Flight* (27 May 1920), p. 567.
65 NA, CO 850/83/14.
66 NA, CO 822/36/6; CO 822/43/17.
67 *The Times* (30 April, 27 August 1938); *Listener* (6 March 1935), pp. 395–6.
68 PDPG (October, November 1934; January 1935; January, March 1937).
69 PDPG (November 1935; October 1936; March, June, July 1937).
70 *Imperial Airways Staff News* (14 December 1935).
71 PDPG (October 1929 and 1930); (January 1931); J. H. Bamberg, *The History of the British Petroleum Company, Vol. 2* (Cambridge, 1994); W. Armstrong, *Pioneer Pilot* (London, 1952).
72 *Star* (12 January 1932); *African Roads and Transport* (October 1938).
73 *Evening Post* (Wellington) (4 May 1937), p. 13; (20 July 1937), p. 10; *Dictionary of New Zealand Biography* (2007); *Evening Post* (Wellington) (27 April 1939), p. 20.
74 Humphreys, 'Ruwenzori'; *East Africa* (31 December 1931).
75 C. Breasted, 'Flying through the ancient Near East', *The Scientific Monthly*, 36 (1933), 62–5; C. Breasted, *Pioneer to the Past* (Chicago, 1943), p. 405.
76 J. Evans, *Time and Chance* (London, 1943), pp. 378, 392; PDPG (April 1932; January 1933); M. S. Drower, *Flinders Petrie* (London, 1985).
77 H. Wayne (ed.), *The Story of a Marriage: Vol. 2, The Letters of Bronislaw Malinowski and Elsie Masson 1920–35* (London, 1995).
78 J. Gladstone, 'Audrey I. Richards, Africanist and humanist', in S. Ardener (ed.), *Persons and Powers of Women in Diverse Cultures* (Oxford, 1992), pp. 13–28; A. Richards, 'From bush to mine', *Geographical Magazine*, 1 (1935), 463–76.
79 H. B. Reynardson, *High Street, Africa* (Edinburgh, 1936).
80 *Imperial Airways Weekly Air News Bulletin* (22 June 1938).
81 *The Times* (13 December 1934; 14 December 1935).
82 *Sphere* (22 April 1935), p. 126; *The Times* (26 March, 22 April 1935).
83 J. Verhoogen, 'We keep house on an active volcano', *National Geographic*, 76 (1939), 511–50.
84 G. Crile, *Skyways to a Jungle Laboratory: an African Adventure* (London, 1937).
85 NA, CO 822/81/15: East Africa Transport Policy Board Report.
86 Cambridge University, Girton College Archive, GCPP Cam 1/5: *Imperial Airways* timetable.

87 E. A. Johnston, *To Organise the Air* (Cranfield, 1995); S. J. G. Hoare, *India by Air* (London, 1927); BL, OIOC, Mss.Eur. C152/11; L/I/1/33; L/PO/1/31(i).
88 *Hull Daily Mail* (8 July 1938).
89 H. Fysh, *The Log of the Astraea* (privately published, 1933).
90 BL, OIOC L/I/1/33.
91 British Postal Museum and Archive (London) (BPMA), PO 33/5367 (2): Report by Sir Christopher Bullock on the Results of his Mission to Africa, 1935.
92 BPMA, PO 33/5525 (1) & (2); 33/4249A (2): Report on Visit to Rome, 1937.
93 NA, CO 822/40/11.
94 NA, AVIA 2/1989.
95 NA, DR 9/68.
96 NA, AVIA 2/1547. On Tillard's findings, see Pirie, *Air Empire*, pp. 224–6.
97 BAAM, N 532: Burchall Papers; *Imperial Airways Weekly Air News Bulletin* (24 August 1939).
98 *Engineer* (15 July 1938), p. 67.
99 *Flight* (18 April 1929); *The Times* (9 May 1931).
100 NA, AVIA 2/401; BPMA, PO 33/4739 (1) & (2).
101 *Cape Times* (11 March 1939), p. 16. Sir John and his wife were guests of Governor General and Lady Duncan in Cape Town before they sailed back to England. Galpin had previously flown to Bahrain on airline business: PDPG (January 1932).
102 G. Coster, *Corsairville: the Lost Domain of the Flying Boat* (London, 2000).
103 BAAM, AW/1/4027.
104 *Imperial Airways Staff News* (January 1934; May, June 1936; March, September 1937).
105 *Imperial Airways Staff News* (3 January 1934), (19 May 1936).
106 E. Bigland, *Awakening to Danger* (London, 1946), p. 138.
107 BAAM, AW/1/3932: Pay and Allowances Bulletin, No. 9, undated.
108 E. Bigland, *The Lake of the Royal Crocodiles* (London, 1939), pp. 21, 287.
109 *Sydney Morning Herald* (18 August 1932).
110 *Imperial Airways Gazette* (August 1934).
111 Hudson and Pettifer, *Diamonds*, p. 79.
112 E. Cobbold, *Kenya: the Land of Illusion* (London, 1935).
113 *Flight* (10 November 1932), p. 1054; *South African Motorist* (January 1933).
114 *Flight* (14 February 1935), p. 184; *Imperial Airways Staff News* (31 August 1935).
115 *Imperial Airways Staff News* (31 August 1935; 12 May 1939); *Imperial Airways Weekly Air News Bulletin* (22 June 1938).
116 *Imperial Airways Staff News* (14 April 1939).
117 *The Times* (31 August 1935).
118 H. E. Turner, *The Fifth Imperial Press Conference* (London, 1935).
119 WAC, S271/1/1; *Singapore Free Press and Mercantile Advertiser* (2 July 1938); R. D. S. Higham, *Britain's Imperial Air Routes, 1918–1939* (London, 1960), p. 235.
120 *Straits Times* (Singapore) (17 July 1938).
121 National Library of Australia (Canberra), MS 5029.
122 *Imperial Airways Weekly News Bulletin* (9 February 1937); *Manchester Guardian* (18 April 1938), p. 8.
123 A. Woollacott, '"All this is the Empire, I told myself": Australian women's voyages "home" and the articulation of colonial whiteness', *American Historical Review*, 102 (1997), 1003–29.
124 Prof. Henrik Larsen (Aalborg University) explained the likely linguistic confusion and confirmed there were no women professors in the Copenhagen Geography Department then.
125 NA, AVIA 2/149. Smoking cabin (before conversion to flight clerk's office and forward mail hold): seven seats; mid-ship cabin: three; promenade cabin: eight; aft cabin: six seats.
126 H. F. King, 'Empire return', *Flight* (15 September 1938), b–e.
127 *Imperial Airways Gazette* (February, April 1932).

128 *Automobile* (South Africa), (September 1932); *Star* (18 January 1933).
129 J. Marston, 'Uganda, land of something new', *National Geographic*, 71 (1937), 109–30.
130 G. H. Pirie, 'Incidental tourism: British imperial air travel in the 1930s', *Journal of Tourism History*, 1 (2009), 49–66.
131 *Flight* (7 February 1935).
132 K. M. Cameron, *Africa on Film* (New York, 1994).
133 *Flight* (27 June 1935), p. 715.
134 *Saturday Review* (17 July 1937), p. 44; *Imperial Airways Gazette* (16 April 1938).
135 J. S. Pudney, *The Seven Skies: a Study of BOAC and its Forerunners since 1919* (London, 1959).
136 H.F. King, 'Empire return', *Flight* (8, 15 September 1938), pp. 198–203, b–e; *Flying* (9 July 1938), p. 6.
137 WAC, microfilmed text of talk by E. Middleton, 10 February 1937; J. C. Earl, 'Achievements of aerotransport', *Chambers's Journal* (1937), 223–5.
138 *Imperial Airways Weekly Air News Bulletin* (6 July 1937).
139 Coster, *Corsairville*; *Aeroplane* (8 June 1938).
140 *Imperial Airways Staff News* (26 May 1936); *Imperial Airways Weekly Air News Bulletin* (26 April 1939).
141 www.gettyimages.com/detail/3312887/Hulton-Archive (accessed August 2010).
142 W. T. C. Berry, *Before the Wind of Change* (Halesworth, 1983).
143 *Lancet* (20 February 1937), p. 475.
144 *Flight* (20 July 1933), p. 730; *Daily Mirror* (18 August 1938), p. 12.
145 *Sphere* (23 April 1933), p. 155; *Imperial Airways Staff News* (17 January 1934).
146 *Flight* (18 October 1934), p. 1103; (17 October 1935).
147 *Imperial Airways Gazette* (June 1935).
148 *Imperial Airways Weekly Air News Bulletin* (22 June 1938), (26 April 1939); British Empire and Commonwealth Museum (London) (BECM), Tape 363.
149 *Imperial Airways Staff News* (3 January 1934; 3 August 1935).
150 *Pathe Gazette* (22 December 1932).

6

Imperial journeys

> *the noise and the speed give you the feeling that you are taking part in a little epic of your own creation, and the sight of the neat ground far down gives you a sense of buoyant superiority over the earth and everything on it.*[1]

As if to re-create the atmosphere of a short stay in a country lodge, *Imperial Airways* kept a 'visitor's book' in its aircraft. Some entries were extracted and reprinted in its smart monthly news magazine. Passengers ('guests') quoted in the *Imperial Airways Gazette* were never rude. Sir Arthur Evans, Rev. J. D. Caldicott from Burma and Brig.-Gen. W. W. Harts, the Military Attaché in the American Embassy at Paris, complimented the airline on its punctuality, efficiency and courtesy.[2] Culling other sources, the *Gazette* carried a message of thanks from Lord George Lloyd after his flight from Africa to London (where he addressed the 1933 Annual General Meeting of the British South Africa Company). In all senses (route, airline, style, preconceptions), flying imperially would have resonated with the ardent imperialism he honed as Governor of Bombay and then High Commissioner in Egypt in the 1920s. As Company Board member, and from mid 1937 as Chairman of the British Council, Lord Lloyd probably flew Empire routes more often than just once again in 1938.[3]

Maj. Grogan, invited by *Imperial* to be its first passenger aboard its first commercial London–Cape Town service in April 1932, had special reason to express pleasure with the trans-Africa air route. His overland travels in Africa had taken two-and-a-half years; *Imperial*'s *Hannibal* carried him across the length of the continent in eight-and-a-half days. As a fifty-seven-year-old air passenger, the Englishman had needed neither his keen sight (Africans nicknamed him 'Leopard Eyes') nor 150 African porters, and the venture did not reduce him to an unkempt, fevered skeleton. Grogan might have reflected on how he might have challenged more effectively for the Newcastle-under-Lyme parliamentary seat in the second 1910 British general election had he been able to travel by air. As it happened, a three-week sea voyage from East Africa had allowed him only one day's campaigning.[4] Twenty-two years later, the *Gazette* gleefully reprinted the telegram Grogan sent to the *Daily Express*. The airway, he cabled, was 'a glorious endeavour of our race'. Perhaps with that in mind, an octogenarian Briton set out on – and completed – her own six-week-long air tour in both directions across the length of Africa with *Imperial* in 1934. It was her first flight.[5]

Care, comfort, convenience, custom

A prominent British businessman in Karachi (proprietor of the city's *Daily Gazette*, Founder Chairman of the Air League of India, and seventeen-times Chairman of the Karachi Chamber of Commerce and Industry) was among those who publicised the comfort and convenience of flying on the Empire routes. Sir Montagu de Pomeroy Webb's account of a flight from Karachi to London in 1932 advised prospective passengers about clothing and reassured them about ambient cabin noise, motion sickness, nervousness and giddiness. With precious condescension, he declared that even ladies would be able to travel as air passengers to India. Coyly, he elaborated that passengers could take their meals and stretch their legs every four hours when planes landed to refuel. The sleeping accommodation was reputedly excellent. The only drawback Sir Montagu pinpointed for women was that *Imperial Airways* did not supply swimming costumes. The sixty-three-year-old was evidently not one of those travellers who pined for shipboard sports and dancing, nor for 'the trip ashore at Port Said for bargains in Chinese shawls and amber necklaces'. Copies of the *Sketch*, *Autocar* and the *Indian State Railways Magazine* in the lounge at Sharjah would have deflected male boredom. And, at journey's end, Sir Montagu wrote, there was something wonderful about the terrace of the Karachi gymkhana and the lounge bar of the London Trocadero being only a few days apart.[6]

An informative account of an *Imperial* flight by an inexperienced air traveller appeared in the *Field* in the summer of 1935. The writer hinted at the need to be alert for sightseeing from the air and during stops. He advised women passengers to lighten their luggage by abandoning crocodile-skin dressing cases lined with silver fittings (just as Lady Hoare had done). Piling reassurance on top of advice, the writer noted that his own initial misgivings about flying to Central Africa were shared by elderly relatives who wondered aloud whether his memorial service would be well attended. In the event, the flight proved safe and comfortable. Female 'voyagers' knitted. Men, of course, read the *Field*.

The airline offered considerate service, said the writer. It shepherded passengers gently and politely, and never dragooned them. Their passports were looked after. There was even no need for tipping. With a touching reverence he argued that as none of the airline's 'native servants' ever asked furtively for money, it must have paid them regularly and fairly.[7]

An *Imperial* ticket bought some comfort and security, but not lightning speed. While the airline bumbled along the world's longest air routes it chose to refer to stately progress rather than pace. However it was disguised, languorous flying was the subject of pointed jests. A report in a 1938 edition of *Time*, the American news magazine, related a fanciful yarn about India's long-delayed independence: the guess was that it might be coming on *Imperial*. Saucier still was a tale about a woman who gave birth on board an aircraft while flying to India. When a flight attendant chided her for boarding the plane, she became indignant. Using language reflecting the stagnant discourse about travel to India, she replied hotly that when she 'got on this ship' she was not pregnant.[8]

Whether or not slow flying was a flaw was contentious. The *Imperial Airways* Chairman, Sir Eric Geddes, was adamant that speed was a secondary consideration for many Empire travellers. In his address to the annual general meeting of *Imperial* shareholders in 1936, for instance, he claimed that a large proportion of passengers did not wish to be whisked over the continents. On the contrary, many wanted to travel in even easier stages to take advantage of leisurely encounters with different places *en route*.[9] As to less tangible elements of Empire air service, courtesy and efficiency were prized, as letters to *The Times* in 1932 show. Passengers also approved the ability to make the best of any situation. When pilots fell ill or aircraft developed technical faults, the opportunity for sightseeing at airline expense was considered part of the service. The presence of English people along the way was rated highly. Many passengers in 1932 may have been relieved that an English couple administered the air station at the Rutbah Wells police post in the Syrian Desert. English women who still clung to the notion of gentlemanliness were said to appreciate the English aircrews and ground staff. They felt they were 'never for a moment out of touch or without the care of Englishmen'.[10]

When abroad, 'Western' could be decidedly different from 'British'. After her circuitous and stressful flight to Kenya as a passenger on a French airline via West Africa, Mrs Tony Spooner was relieved to sink into *Imperial*'s armchairs for a Rhodesia–Nairobi flight. She boarded at Broken Hill after crossing the Sahara to Kano, and making an emergency landing in the Cameroons where the French aircraft was encircled by

Africans and where her anger welled up at 'hopeless disorganisation'.[11] 'Westernisation' certainly had benefits, however. After the start of the Croydon to Cape Town air service in January 1932, a British newspaper editorial remarked that the adventure, with all its changes of aircraft, sounded arduous, but that places *en route* had been 'occidentalised to the point of French cookery and American cocktails'. Everyone who had tested the service in its experimental stages was said to have noted the comfort of each night's rest in the heart of darkest Africa: 'one may sleep as comfortably in Juba as within the sound of Big Ben'.[12]

Even if the pace and frequency of *Imperial*'s commercial service did not please users, its regularity did. A correspondent in South Africa's *Cape Times* drew on G. K. Chesterton to convey admiration: 'it is not the isolated instance, like a man without a nose, that is wonderful; the miracle is in the thing that is endlessly repeated, like the fact that everybody has a nose'. Air services were beginning to choreograph imperialism, especially in places not yet touched by mechanised land transport. Slow, low-flying aircraft passing overhead were like a remote clock chime or bugle. At a Rotary lunch in the Hotel Russell in London's West End in October 1933, Sir Samuel Instone told his audience that government officials in Africa were finding that passing aircraft were useful in verifying statements made by Africans about the day or time of other occurrences. Whereas they were 'usually very vague as to time, which meant little to them', they now recalled the timing of events in relation to whether they were before or after an aeroplane passed. Was it just a slip of the colonial tongue that Instone noted in the same breath that even wild animals were growing accustomed to aircraft passing overhead, and no longer stampeded? Might hungry Africans one day also emulate elephants and stand waiting for buns to drop from passing aircraft, like manna from heaven?[13]

Imperial's efficiency and care for passenger comfort counted a great deal, but so did its response to incidents when that regularity was compromised. In a letter to *The Times* in 1932, a Mr G. Kayser, a Hertfordshire resident, wrote admiringly of his flight from Salisbury, Rhodesia. On arriving at Dodoma (Tanganyika), the pilot was so ill with fever that a doctor refused to allow him to proceed. Within five hours a relief pilot arrived by air from Nairobi. He flew the passengers 400 miles back to Nairobi, landing at night and enabling the air mail to leave on schedule the next morning. There was engine trouble again just as the aeroplane was leaving Aswan, but in trying conditions the pilot managed to get back to the aerodrome.

While arrangements were being made to get the passengers to Alexandria by train, they took a taxi to the Temple of Philae. Back in Aswan they found that the Egyptian railway authorities had refused to accept the pilot's signature for rail tickets. In addition, the National Bank of Egypt blocked him drawing money and would not allow *Imperial*'s ground representative to overdraw his private account, even though he frequently presented the Company's official cheques. An expatriate Englishman working on the Aswan Dam (who had rescued the sightseers when their taxi's tyre burst) stepped into the breach again and gave the pilot a £60 cheque drawn on his private bank account. After leaving Alexandria a day late, the passengers reached Brindisi in one day and were still able to catch a train that arrived in Paris at the correct time.

It was *Imperial Airways*, Kayser wrote somewhat misleadingly, that 'triumphed over all obstacles and landed the mail at the scheduled time'. His journey had been 'remarkably interesting, embracing, as it did, such a variety of conditions, climate, and scenery'. It was, indeed, 'a unique experience and, quite apart from the immense saving in time, a voyage well worth undertaking'.[14] In Kayser's case the delights of air travel clearly compensated for the frustration. Setbacks could be eased by prompt action. In 1934, when a *Daily Mail* columnist *en route* from London to Singapore lost all his luggage (the Italian launch sank when carrying luggage from the Brindisi shore to the flying boat), *Imperial Airways* officials at Athens, the next stop, bought him a fresh outfit of clothing to replace the garments lost overboard.[15]

Disruption

Air service frequency was less of an issue than safety. Aged sixty-nine, Sir Philip Richardson (a British marksman and ex-M. P. who had flown the round journey between Britain and Kenya three times by 1934) spoke defensively about aviation's safety record. He noted that 6,000 people died annually in motor accidents in Britain in the early 1930s.[16] A reassuring perspective was welcome because the safety record of the Empire airline was neither flawless nor invisible. Photographs published in a twice-printed 1933 book by a prominent ex-parliamentarian and established commentator would have annoyed *Imperial*: the most intact feature of the wreck of the *City of Karachi* from which he escaped unhurt at Gaza in 1930 was the airline name painted in large lettering on the fuselage. The photograph in a 1935 edition of the *Sphere* must have been unsettling: it showed the *City of Jodhpur* which, having suffered engine failure, ended up in a Ugandan swamp ten miles from Entebbe. A road had to be built to salvage the plane.[17]

Following an unscheduled stop it must have been comforting to discover seamless British imperial hospitality. In March 1934 when an *Imperial* aircraft was damaged by a sudden squall while refuelling at Kuwait, the local British Political Agent hosted passengers for twenty-four hours while a replacement aircraft was awaited. The aircrew, which had to wait longer for spares from London, received food, blankets and baths in a large tent that the agent ordered erected on the aerodrome. There was still no rest house or passenger shelter at Kuwait three months later when *Imperial*'s *Helena* put down unexpectedly in poor weather. Eight passengers were given food, beds and bedding at a house belonging to the Anglo-Persian Oil Company.

Even British citizens who flew on the rival Dutch airline could expect care when it counted. After a KLM crash at Bushire in 1935, the British Residency opened its doors to two of the six passengers who were English – a soldier in the Twentieth Lancers and the principal matron of the Northern Command in India. In 1936, when *Imperial*'s *Horsa* landed prematurely on the Arabian mainland, the Political Agent at Bahrain initiated an RAF search. The eight passengers (including the wife of a Melbourne stock-broker) and crew of four were found and taken to Bahrain the day after the landing. In temperatures reaching 130°F they survived for thirty hours off a few sandwiches, a small bottle of water, chocolates, a dash of whisky and dew collected in sponges off the aircraft wings. Water was rationed to less than a full cup three times a day. Thirst made eating difficult. A stay at Bahrain supervised by the British representative would have settled nerves more than the air-drop of letters asking the Al Mirrah Sheik to guard a piece of Empire hardware. Doing so nevertheless enabled *Horsa* to be recovered intact and repaired in situ. *Imperial* rewarded the 'tribe' suitably.[18]

Presents were not always accepted. The Emir of Hasa declined gifts after the rescue of *Hanno* in April 1933 when the *Imperial* captain landed prematurely in a prohibited flying area seventy miles west of Bahrain. Local people took him and the two passengers food and a tent. Next morning, they brought spades, ropes, and strips of matting and wood to prise the aircraft out of soft sand.[19]

Gracious local assistance had also been offered after the emergency landing made by *City of Glasgow* in December 1932 in the Nubian Desert between Cairo and Khartoum. According to one of the passengers, Mr E. Kentish Barnes, a farmer returning from London by air to his Kenyan estate, the twelve passengers spent a day and a night on the ground before being rescued by an *Imperial* relief plane sent from Khartoum. At night they kept warm under covers removed from the aircraft seats. The half-bottle of cognac in the aircraft's emergency ration box helped. Some Sudanese railway officers who happened to be inspecting a part of the line in the vicinity gave up their train saloon for the four women passengers.[20] Seventeen passengers of *Imperial*'s flying boat *City of Stonehaven*, which was forced down by engine trouble in a remote district on the Zeraf River in 1932, reached Malakal none the worse for their adventure. Their only complaint was that the relief aeroplane took canned peaches and cream instead of bully beef and bread.[21]

Publication of photographs of *Imperial*'s dead duck *Corsair* flying boat stranded in a remote corner of the Belgian Congo in 1939 occurred after wartime cessation of civil aviation and could not damage the carrier's reputation. Nevertheless, the forced landing was a reminder that Africa was untamed. An official account of the incident blamed atrocious African weather conditions. In fact, the misfortune followed defective fitting of a new navigational compass at Entebbe. An obscured rock in a strange river played its part. Fortuitously, the Belgian District Commissioner was on hand to attend to the thirteen passengers (including an American couple on honeymoon and a relative of Prime Minister Chamberlain). None was hurt. One account of the events has the District Commissioner feeding the passengers and driving them to a small border hotel fifty miles away, from where they proceeded to Juba overland where a 'Greek' storekeeper provided for their needs (except for ladies underwear, for which there was said to be no local demand).[22] Geoffrey Pett, then *Imperial* station manager at Juba, recalled events differently. After receiving radio messages about *Corsair* landing at Faradje, he and his coxswain drove the *Imperial* station wagon, and one belonging to the Juba Hotel, 100 miles to rescue the passengers. They all returned on a five-hour road trip with luggage and mail aboard congested vehicles to the Juba Hotel (where Pett rented two rooms, one as the *Imperial* office and the other as a staff restroom).[23] Warily, surely, the passengers boarded the next *Imperial* flying boat service northwards.

Much slower rescue was part of the scenario sketched by a writer nine years previously. *Imperial*'s *Gazette* did not carry an extract of his article, which, after all, was directed at private pilots rather than airline passengers. Africa would not distinguish between them. The article amounted to a survival manual for air crash victims. Most passengers would have been alarmed by the prospect of coming down in remote, arid scrub, or in a vast marsh, and having to endure scorching heat, 'murderous nomads' and frightening insects and beasts while waiting for a rescue party. The chances were small of an aeroplane in trouble being able to land on a prepared strip near settlement in the vast continent. Even then, as the Duchess of Bedford found out, the list of helpful contacts in the pilot's manual might be out of date: the local doctor at Nimule landing ground on the Sudan–Uganda border had been gone two years when she and Capt. Barnard sought him out in 1930.[24]

The hazards of flying would have been self-evident to passengers on the *Imperial* aircraft indirectly involved in an air rescue in 1937. They, and readers of the subsequent reports and listeners to the BBC radio broadcast, would have been unnerved by the sight of either a smoke distress signal or a mirror reflection from Brig.-Gen. A. C. Lewin and his wife, sole occupants of their private plane that was lying on its back in a swampy morass in the 'mosquito-infested' Sudd. Waylaid by a defective compass and out of fuel, they were forced to land while flying to their home in Kenya. They survived a *haboub* storm, but it was five days before they were spotted by an *Imperial* pilot, who the next day led eight RAF planes from Malakal to the site. The RAF dropped provisions for five further days. Some sank into the mud. A party of local police and Dinka men who barged and tramped seventy miles finally rescued the couple ten days after their forced landing, stretchering out Mrs Lewin.[25] Rescue came too late for the unnamed 'native boy' who fell off a flying boat refuelling barge at Wadi Halfa and was killed by a crocodile.[26] The tragedy was almost certainly kept from passengers, who were unlikely to have been aboard during refuelling. The human cost of providing imperial transport – sea, rail and air – was seldom transparent to its users.

After an emergency landing in 1938, the *Ceres* flying boat (carrying four passengers, a crew of five, and 1.5 tons of mail and freight) spent eight days on Lake Dugari (between Raj Samand and Gwalior). The presence of the 'great white bird' amazed the local population: priests with gaudily painted faces, tribesmen, women with babies on their backs and naked boys gathered round 'in excited wonder'. At first they refused to assist for fear of crocodiles, but after Capt. Gurney shot two of the beasts he won the people's confidence. It took several days to survey the Lake depths (for take off) using a raft of banana leaves and planks. Initially, thunderstorms prevented wireless communication but, once located, other flying boats dropped supplies of food and water by parachute. In due course, Mr Mukerji, the Station Superintendent at Gwalior, brought a relief party; they had driven for nineteen hours across flooded and bogged roads. Taken ashore, the passengers had a memorable experience in a picturesque and historic place: 'the Natives performed (with home-made whistles and tom-toms), issued invitations to go tiger shooting, and even pressed offers of marriage on one of the passengers'.[27]

Disservice

Writing which consoled, warned or complimented was supplemented by complaints about the experience of flying on the Empire routes. As an *Imperial Airways* passenger, the Duke of Norfolk made his mark by complaining about the mean accommodation he was assigned in a Luxor hotel. Even before *Imperial* began operating to the East, an

Anglo-Persian Oil Company employee judged airliner seating too uncomfortable for a ten-hour journey.[28] Not all Englishmen were patriotic flyers, and they protested by buying foreign airline tickets. In 1936 Jim Mollison used KLM to return to London from Singapore because the airline offered a faster, quieter service at a fare lower than *Imperial*'s £156. His decision attracted critical comment in the British press.[29]

When visiting Australia in 1936, Mr W. H. Pilkington, proprietor of the St Helen's glass firm, patronised KLM between London and Singapore, and Qantas Empire Airways to and from Brisbane. Although he thought it an exaggeration for an Australian passenger on his return flight to tell an *Imperial Airways* captain at Gaza that the British service was no use, he nevertheless made several asides in his flight diary about the relative slowness of the British airline. When passing an *Imperial* airliner in the air he noted the illusion it gave of going backwards. On the outward journey Pilkington arrived ahead of the *Imperial* aircraft that had left Croydon three days before KLM. The Empire airline, he wrote, had lost a lot of prestige 'out East'.

So had KLM, according to one British passenger. He complained to KLM's (Indonesian) station manager at Amsterdam's Schipol airport that both the co-pilot and steward on the flight from the Far East had been Indonesian, and that he had to spend his evenings at the hotels and rest houses *en route* in close company with non-Europeans.[30] Pilkington let pass his sharing an aircraft cabin with a Parsee and an Anglo-Indian lady, but he did record responses to colonised peoples and places. Hot, dirty, smelly Baghdad had him exclaim, 'if the East is all like this, give me the West'. 'A clean hotel in the Near East makes all the difference to a European traveller', he insisted. A bird's eye view of the riverine Garden of Eden seemed to him just ugly. Palestine was surprisingly small. A chemical plant spoilt the Dead Sea shore. Flying over India he was struck by the monotony of the landscape, the ugliness of Karachi and the pathetic sight of the *R101* airship mooring mast still standing six years after it became redundant. After an hour in a hut near Koepang aerodrome he vented the view that the Timorese were 'exceedingly primitive, stupid and slow'. Even his Australian fellow traveller fell short of English values: having done all the sights in 'second-rate' Oriental cities he elected to go for a haircut in Athens rather than see the hilltop antiquity which he termed the 'Pantechnicon'. After encounters with 'philistines', zealous customs officers, flies, insects and heat, Pilkington looked forward to England.[31] Flight, it appears, did not automatically generate international tolerance.

Whether by choice or circumstance, Lord Moyne and Lady Broughton also flew on Dutch aircraft. Their effusive letter of appreciation must have delighted KLM. Its originating address carried the full weight of British imperial authority: 'The Office of the Comptroller, His Excellency the Commander-in-Chief's Household, New Delhi'.[32] Many southern Africans must have wished for a choice of airline after successive service delays and interruptions. In 1932, a disillusioned South African reckoned that *Imperial* was unlikely to have its passenger contract renewed in Central and Southern Africa. Referring to 'chaos' there, the writer noted that *Imperial*'s pilots had undeservedly pulled it through 'the worst managed and most impossible show', one that 'would be a roaring farce if it were not a tragedy'.[33]

Roaring over Africa

Winifred Vergette returned to London by air from Central Africa after spending the 1931 northern hemisphere winter there with friends. She contrasted her seven-day flight home with (what she said was) the similarly priced but much longer twenty-five-day outbound sea voyage, followed by a three-day train journey from Mombasa. Her *Imperial Airways* journey, she said, was 'one of the most delightful trips' she had ever made. She had never flown before, and her friends thought it rather risky. Undeterred by illness which made her postpone her flight three times, she was not perfectly recovered when she eventually left Port Bell on Lake Victoria, but she 'arrived in England a well woman'.

From her home on the Mall, Kensington, Winifred wrote enthusiastically to the editor of *The Times*, noting that the flying boat had descended to show passengers elephants and giraffe. There had been opportunity to saunter along the Nile at Wadi Halfa, enjoy tea on a motor yacht in the Mediterranean and take a motor trip up a mountain at Corfu. Her verdict of care noted nights at the best hotels *en route* and at delightful rest houses in the desert, excellent food everywhere and hotel luncheon baskets where necessary. It was 'a real joy' for a woman travelling alone not to have to arrange rooms or meals. It was also a relief not to tackle luggage and struggle with it through Customs: 'wherever we landed cars were waiting to take us to the hotels with our baggage'.[34]

The impression that flight made on airline passengers went beyond approval or disapproval. Air journeys were an experience, but they were also meaningful: 'flying the Empire routes gave many Britons the same proprietal sensations they had so long enjoyed from their [ocean] liners'. Expounding her claim, the imperial historian Jan Morris cited Sir Edward Buck's account of his flight north across Africa in the company of Lord and Lady Chesham (who had interests in Swaziland) and Mr Kanthack (Companion of the Order of St Michael and St George, formerly of the Indian Irrigation Department). 'I was an

Imperial passenger, a proud title', Buck declared. At Entebbe, he visited Government House briefly, and congratulated the Governor on his new appointment to Nigeria. They shared some Indian recollections. At Juba, where the passengers spent a night, they saw that 'the paragons of the Sudan Service' had built a model village in the bush. It featured a mission church, a hostel ('clean') and native housing ('well planned'). Even the District Commissioner was spick and span. He rose courteously from his desk to greet his evening visitors.

During a short stop of a couple of hours in Khartoum, an army major showed the group around the zoological gardens. Next day they spotted the Aswan Dam, 'the Empire's pride'. When they landed at Luxor for the night, Lady Chesham insisted on visiting Tutankhamen's tomb, a replica of which she remembered seeing at the 1924 Empire exhibition at Wembley. On the fourth day the four 'descended thrillingly' over the pyramids to Cairo, where the Union Jack 'billowed serenely'. The High Commission gardens could be glimpsed beside the river. At Heliopolis, a line of uniformed *Imperial Airways* officials awaited them, snapping into a salute as if piping them ashore from an imperial warship.[35]

The imperial trappings of flight seduced Sir Edward. Others also reclined comfortably into the clear blue and fluffy white topography of the air Empire. Writing in 1935, William Makin was reassured by flying into Entebbe over the golf course, the government house with its swimming pool, the pretentious white-painted bungalows, the fine red roads and the parade ground for troops. Beneath the clouds were 'all the appurtenances of British administration in the tropics'. He saw at a glance the Union Jack formation of Khartoum's avenues and streets; the layout seemed to him 'the very embodiment of British Imperialism in Africa'.[36]

In similar vein, a young British colonial officer flying back to his Bulawayo posting in 1936 let his mother know about the design hallmarks of Empire. Nairobi was well laid out and looked very nice from the air, he wrote. It was 'a real feather in the cap of the British'. He identified hill houses, tennis courts galore, three games of cricket in progress, three large well-appointed hotels, Government House, the cathedral, and a spaciously laid out railway station with a platform wide enough to parade six regiments. Mother would probably have been comforted to know that her son ate a good English breakfast at Entebbe and did so in appropriate company. The young man knew enough to tell her that the English people among whom he sat were smartly attired. Standards and style were important: the men were shaved and wore shorts and helmets; the women were dressed in cool white frocks.[37]

British oases were preferable to Makin than what he diagnosed as Africa's Dantesque landscapes of jungle, creeper, insects, mud, slime, reptiles and fever. He also shrank from 'the devil tattoo of the Dark

Continent', namely the cavalcade of game, slave trails and beating drums. Like Makin, another visitor to Khartoum, Grace Crile, found solace there. She strolled along the quay, passing a general's fenced mansion, a lieutenant's Italian villa and the Governor's house. They resembled palaces, she wrote. Like anchored or berthed steamers, the presence of impressive British aircraft at Heliopolis was also power dressing: 'England needs to look powerful in Egypt.' The effect would be enhanced on days designated for 'dressing ship'. Lines of flags had to be flown along the length of *Imperial Airways* aircraft to mark the anniversary of the King's accession and coronation, his and the Queen's birthday and Empire Day. British airliners in Egypt on Independence Day and on King Fuad's birthday were decorated similarly.[38]

Imperial flying had a less robust face too, being part of the domestication of Empire. *Imperial*'s refuelling and refreshment stops, as well as overnight stops, were little Englands. They did not start that way, as the British Resident in Sudan's Darfur province noted. When *Imperial* sent out three senior officials to survey Geneina, where a night stop was planned, the typography on their maps denoted a place the size of Cairo, Khartoum or Paris. He had to tell them that there was no hotel, no electric light, no power and no piped water.

Neither under-provision nor rusticity struck the Criles when they stopped for tea at Kareima in the Sudan on their 1936/37 African flight. Places were set out at a table covered with a linen cloth 'as white as natives' teeth'. For food, there were middle-class English favourites: Huntley and Palmer's biscuits, tea, cake and orange marmalade. Later, at Malakal, there was a more substantial meal of Nile fish, veal, potatoes, beans, compote of mixed canned fruit and coffee, served on pretty china embossed with the airline's name. 'This is England – off at the end of nowhere.' The service was familiar and correct. In Egypt, *Imperial*'s local employees dressed in clean linen. They wore coloured waistband and turban. Their manners were 'dignified and pleasing', befitting wage earners. One could be proud of them. They were a far cry from local Bedouins, who dressed either scantily or shabbily in yards of dirty flapping cloth, and who passed the time shuffling alongside disconsolate donkeys and mongrel dogs.[39]

Service was everything, especially in modest surrounds. At Aswan, where passengers lunched in a tent, the floor of small stones was watered frequently to create a cool atmosphere. Facilities at Wadi Halfa were better. Resting in a building reputedly occupied by Gen. Gordon and Lord Kitchener would have made ardent imperialists glow with pride and privilege. Arrangements were made to admit passengers to the local British Club, where they could read newspapers and consult the latest news telegrams. At Entebbe, 'Empire Passengers' were hosted in a small

Figure 8 Little Englands: aircraft and rest stop at Entebbe (Uganda), 1936.

guesthouse whose porch was draped with colourful bougainvillaea and roses. A simulated English country home lay beyond the make-believe English country garden. Inside, there were decanters of orange juice, dishes of tiny bananas, slices of juicy pineapple and luscious papaws.

At Juba there was a modest hotel in the vicinity of a minor military base, some Indian stores, a wharf and a few native huts. Facilities at Dodoma were 'clean and civilised'; refreshments arranged delightfully on a snow-white tablecloth showed that some English woman was doing her bit to make the tropics better and brighter. Enterprising Africans living close to airfields brightened their own lives by tailoring trousers out of windsocks, at least until the replacements were sewn out of convict dress.[40] Perceived African disrespect for Crown property abroad, and for the safety of those who depended on even the simplest flying aids on the ground, might have sharpened clamour for firm imperial rule and instruction.

Whimsy percolated an article published in 1939 in the *Journal of the Royal African Society* about 'the romance' of the air mail to Africa. The author, Robert Brenard, resigned that year from his public relations position with *Imperial*. His London-based perspective, and imperial habit, had him title the piece as if the air mail traffic was one-way. His overview of British imperial aviation on the continent told anecdotally about iced drinks served by stewards at the rest house at isolated Mbeya and about Africans gawking at aeroplanes in flight. Were the big birds covered in feathers, and could their eggs be taken back to the village? At Mpika, African 'boys' were given the task of lighting the landing ground. By rattling their flare buckets they scared off lions drawn to the landing ground by the presence of oxen used to haul a grass mower. The engineer-in-charge at Mpika carried a pistol that he could use to frighten the lions and chivvy the reluctant Africans.[41]

Text and photographs about air travel were in the tradition of chronicles of early overland expeditions. A travelogue which the *Geographical Journal* published concerning a flight from Croydon to Kenya in 1931 was reminiscent of accounts printed by the Royal Geographical Society fifty years previously, although the sequence of impressions on a flight came faster and were more jumbled in the telling. After the thrill of taking off in a flying boat at Brindisi, the adventurer's diary filled with notes about lunch aboard the *Imperial* wireless and depot ship 'Imperia' at Mirabella Bay in Crete, the British Boat Club at Alexandria, and a glimpse of Luxor and Abu Simbel. To the south, the vast swathe of Sudd marshland caught the eye. Other notable incidents and sights included wake-up calls at 3.45 a.m. after 'overnight' rest stops, herds of elephant and giraffe, a hippopotamus swimming close, a pair of mislaid field glasses, and a boy who fell into the water and had to be 'harpooned'.[42] Strains of adventure permeate Grace Crile's account of a flight to and from Kenya in 1936/37. Like the landscape, the pages teem with giraffe, elephant, rhino, lion, eland, hartebeest, gazelle, hyena and warthog. The game, and the storks and herons, vie for page space with Abu Simbel, the Temple of Isis, Karnak, the Giza pyramids, the Temple of Thebes and the Colossi of Memnon.[43]

Other flight diaries contained similar lists. On a July 1933 flight from Croydon to Cape Town, Alex Barnes recorded the moonlight taxi trip to the pyramids ending at 11 p.m. and the 1.30 a.m. wake-up call for departure at 3 a.m. (it was easier to get airborne before the sun's heat rarified the air). The £130 fare purchased seventy-two hours in the air, 1,273 miles by rail, thirty-one stops and four changes of aircraft. Among the compensations for the regimen were airline staff who were 'the most courteous of Englishmen it is possible to find'. And, over Uganda, the pilot circled so passengers could see hundreds of elephants stampeding from the plane's shadow and engine noise. Lady Cobbold's narrative of aerial sights and flying experiences in Africa in 1934 was similar. Her qualified enthusiasm for flying itself did not overshadow praise for *Imperial*'s efforts.[44] On another flight, the two passengers who jokingly threatened to sue *Imperial* for breach of contract (publicity promised views of wild game) would have been pleased by the aeroplane commander's flying over numerous herds of buck south of Mbeya the next day.[45]

In their efforts to describe flying, passengers dealt in fact, fiction and interpretation with varying degrees of wariness. In the new *Geographical Magazine* the plainest diet of history-from-aloft was peppered with aerial photographs – the new three-dimensional mapping – taken from an *Imperial* flight to India.[46] *Chambers's Journal*, like Cobbold's account, carried spicier text about Malakal's nearly naked Shilluk men and their curious head hair. They were indeed a reminder that London was far away. It was mistaken to write, however, that a river blocked with floating Sudd islands, and thick with hippopotami and crocodiles, was 'a matter of indifference to a flying boat'. It was stretching credulity to remark that when crossing the equator the *Imperial* airliner flew 'very slowly to celebrate the occasion'.[47] In 1939, if not before, the airline did mark the crossing by giving each passenger a large 'Certificate of Contemporary Travel' signed by the flying boat commander. The Assistant-Director of Works at Malakal received one of the colour-printed documents. A Royal College of Arts graduate illustrated the document with artwork suggesting passage over a tropical Eden. Its illustrious provenance did not tempt the public servant to frame and hang his trophy.[48] A Nile air service, British or not, made no impression on another public servant based in the southern Sudan. Then in his twenties, Wilfrid Thesiger avoided intrusive mechanised transport when travelling in North Africa. Even when taking up his appointment in the Sudan Political Service in the 1930s, or going on leave, he travelled to and from Malakal by river boat.[49]

Exaggeration about flying came as easily to some as minimalism came to others. Statistical details (cruising altitude and speeds, departure times and delays, passenger numbers) fascinated some passengers. On his 1933 flight from Johannesburg to London, Mr G. Armstrong noted these details carefully alongside comment on air pockets, mist, motor car ferries to hotels, a visit to a coffee estate near Nairobi and descending to see animals.[50] The barrister H. C. Willis wrote with exemplary detachment. His lengthy and meticulous dossiers on three *Imperial* flights across Africa are filled with objective numerical information about times of arrival and departure, distance and speed. He treated weather and landscape as matters of factual scientific record.

Sixty years on, the dull detail of Willis's 1938 flying boat itinerary sheds light on flexibility untroubled by considerations of tight landing slots, restricted flight paths and complaining passengers. There was a long lunch ashore at Alexandria, where a pilot had family; the aircraft deviated off course briefly to offer a better view of an elephant herd; at Kisumu Willis was awoken at 5.15 a.m. to make an unexpectedly early start following reassignment of aircraft. He only just noted the embarrassment of the most junior member of the station staff who was sent to rouse him. Willis rarely conceded to having impressions. He lapsed once when admitting how at Dar es Salaam harbour he had felt 'delightfully and snobbishly superior' when passing the passengers who were coming ashore from the British India boat. His puffed head got the better of him. On the basis of appearance rather than firm evidence he wrote after take-off that it seemed to be a favourite game of *Imperial*'s flying boat pilots to deliberately frighten and shower local boatmen.[51]

On both the African and Indian services, *Imperial*'s passengers were pampered from Cairo onwards. There, the airline used Shepheard's Hotel and the Continental Savoy Hotel. They were said to be among the finest in the world. Shepheard's was legendary, the epitome of style and sensation, a playground and watering hole for the aristocracy, dashing officers, brave explorers and gorgeous film stars. Sir Richard Burton, Henry Morton Stanley, Gen. Gordon and Lawrence of Arabia had been guests. Ross Smith spent the night there in 1919 during his pioneering flight to Australia. Sir Samuel Hoare stayed four nights during his RAF flight to the Middle East in April 1925.[52] Adding Beryl Markham's name to the list of intrepid adventurers and conquerors would have led to gasps and gossip. She seduced an *Imperial* captain in the Hotel in 1934, but possibly not in 1938 when she passed through as one of the airline's passengers *en route* to London from Cape Town via Kenya. In earlier years, many of *Imperial*'s passengers sleeping overnight at Shepheard's must have felt they were the new agents of Empire. Pungent associations were in the air elsewhere too. A sentence buried in a 1932 magazine article promoting winter touring to Kenya played on fine imperial history: 'to see people leaving [Lake Victoria] by air for Croydon, and think of Speke and Burton, of Baker and Stanley – this is indeed romance writ large'.[53]

The writer Elspeth Huxley (resident in Kenya for a dozen years until 1925 and once employed as Assistant Press Officer to the Empire Marketing Board) recalled a six-and-a-half-day *Imperial* flight from Entebbe to Croydon in 1933. Her husband booked their flight, undeterred by his own recent experience of being stranded in the Northern Rhodesia bush for three days with the crew and passengers of an *Imperial* plane that had got lost and run out of fuel. The sense of adventure was palpable. The discomfort left a lasting impression on twenty-six-year-old Elspeth. The journey was bumpy, hot and stuffy. Passengers were airsick. The midnight temperature at Khartoum was 110°F. A 'repellent' breakfast was served before sunrise at Atbara. At Wadi Halfa an inedible lunch smothered by flies was set out in a stiflingly hot tent. Passengers endured a night in a horrible hotel at

Aswan. On the positive side there was pleasurable sightseeing (the Tomb of the Kings, the Khartoum zoo), and evening drinks made all the more delightful by the crew of 'five young men, all cheerful and friendly'. At Alexandria the passengers were woken at 3.30 a.m. It was reward enough to reach Corfu as early as possible: the sight of the Royal Navy's Mediterranean fleet at anchor restored a sense of ordered normality.[54]

It was not this reality of flying that a fourteen-year-old boy photographed on the ground at Kisumu aerodrome in 1933. Stationary, and beyond his use, *Imperial*'s three-engined *Hercules* radiated sheer romance. The two smartly dressed White women shown striding toward the bi-plane may have been under that same spell. Lovingly reprinted in the boy's life history sixty-four years later, the caption to the image muddied reality in another way: the 'Javelin' aircraft type he remembers photographing never existed.[55]

Elspeth Huxley's account was manifestly not scripted by or for *Imperial*. Neither was her mother's recollection of a flight from Kenya to England and back in 1935. Published in the 1980s, Nellie Grant's letters recall an incident-packed return journey variously in the company of a Yale University professor, a Norwegian journalist, a German, a Belgian, two Italian diplomats, an Indian and two Southern Rhodesian youths. Sand whirlwinds forced a landing in poor visibility at Kareima, 'a tiny place with only one European, the wireless man'. The plane was badly damaged. Engineers and parts were summoned from Cairo. Attempts were made to repair the torn wing fabric with the wireless operator's shirt and the windsock. The steward ('marvellous . . . a perfect man') commandeered a donkey and galloped off to a village to get food. Next day a relief plane took the passengers in two batches (without luggage) to Khartoum. There they tried, unsuccessfully, to board *Imperial*'s next scheduled southbound service. Only the mails were loaded. Making their way south on their repaired aircraft, the passengers discovered that the mail-carrying aircraft had crashed at Entebbe.[56]

Viewing Asia and India

Like the African airway, the new air route to India was an irresistible subject for writers. Two of the first three expositions in 1929 were by women. An American wrote up her eight-day September flight from Karachi to Croydon in the company of a lone male passenger. Her photographs gave a feel for the aircraft, its pilots, the historic and desolate landscapes, and the primitive facilities *en route*.[57] Mrs Kingdon Ward (wife of the eminent English botanist explorer) published her account of a week-long flight from England to India in an edition of the *Empire*

Review. She made her journey as the quickest way of reaching her husband, who was critically ill in Indo-China. Her flight from Croydon to Basle, and then from Genoa to Alexandria by flying boat, proceeded without incident. Her onward journey to Karachi was more eventful, and included a desert rescue. Arthur Eggar's account drew attention to the sentry at Rutbah Wells, whom he imagined protecting the two women and three child passengers from abduction and enslavement by wandering tribesmen. He, rather than the two women writers, was affronted by the way spotless linens packed in London bore the mark of many customs officers' hands.[58]

Alarm was never planned, but the novelty, sensations and sights meant that a flight to India could be the experience of a lifetime. That, anyway, was the opinion of twenty-five-year-old Robert Byron, who flew for the first time when travelling on *Imperial Airways* to Karachi. Lord Beaverbrook paid his passage in August 1929 in exchange for four articles about the new England–India air mail service for his *Daily Express*. The eccentric, ill-tempered member of a generation of gifted British travel writers grumbled about the indignity of passengers being weighed personally and the limitation of a 266 pound weight allowance (person plus luggage). The ambient noise and the cramped space were irritating too.[59]

It was unlikely that restrictive air travel would appeal unreservedly to the flagrant individualism and eccentricity of people like Byron. Yet he claimed to be profoundly refreshed by his journey above deserts, through fiery air, via a lonely fort full of human bones, across stunning mountain ranges, pausing for a meal in a tent in Baluchistan half-a-day's camel ride from the nearest habitation. Byron was pleased by a succession of warm and efficient welcomes. Among others, they avoided exasperating formalities at sea ports. Other appreciative passengers acknowledged 'British genius'. The Persians were not taken in. Whereas they had used goat's blood to propitiate aircraft when they first arrived, Byron found that officious searches were the new routine. His annoyance was aggravated after his pleasant break in Iraq. At the Nairn Motor Transport Company's hotel at Baghdad, he sat in a cool lounge containing 'every known English periodical'. At a cabaret in a city nightclub, he spotted Arab gigolos, fierce Bedouins and buttoned-up Englishmen trying to appear indiscreet.

Landing at Karachi triggered the first of many contradictory feelings that affected Byron with growing intensity the longer he stayed in India. His journey there had been by air, but thereafter his experience was not unlike E. M. Forster's, whose passage had been by sea; the means of arrival left the destination untouched. Byron's first sight was a string of camels walking along an asphalt road. Alongside there were English signposts; in the background were English church spires. Flight

Figure 9 Rutbah Wells (Iraq), desert track to Baghdad and landing ground, c.1936.

had neither produced nor ruined that pastoral scene. One could merely enjoy the familiar more quickly having sampled exoticism briefly.

Byron was attracted by arrangements that retained the best of the past while also admitting innovation. On his flight he discerned an emergent 'fraternity of the air' comparable to that of the merchant marine. In its new guise, he presumably hoped that this ancient spirit would serve the Empire well. He wrote explicitly of his hope that aircraft might be the equal of ships and 'the honour of ships' in their contribution to the cohesion of the British Empire. It was a moot point whether international or Indian domestic aviation would overturn some of the less honourable features of British India, which both Forster and Byron experienced and wrote about. It was while travelling 16,000 miles by rail in India that Byron encountered 'an appalling spectacle of British snobbery, race prejudice, complacency, and imperiousness'. Some of these traits may have manifested at sea: around the end of the First World War a P&O steamship skipper declared that British expatriates in India were the most shocking of all. Relatively high airfares, and on-board segregation of pilots and passengers, may have shielded *Imperial*'s captains from any such colonial excesses.[60]

Arrival in India stirred emotions, and so too did the journey. Some writers showed this, others chose not to. Sir Philip Sassoon, an unworldly, un-English, Baghdadi figure, clung to his public mask. The junior Air Minister, a sensitive aesthete always in flight from his Jewish roots, did not betray his past when writing about his solo flight via the Middle East to India in 1929. Privately, however, Sassoon might have reflected that exactly 100 years earlier his great-grandfather began his own flight overland from the tyrannical ruler of Baghdad. David, founder of the Sassoon dynasty (and of Philip's fortune) took thirteen arduous years to reach Basra; his great-grandson took three days.[61]

The Middle East air route had the capacity to evoke private histories as well as grand history, biblical or otherwise. Ross Smith felt the tug of the past during (or after) his 1919 Australian flight. Well read, well researched or well ghosted, his report was richly reflective. It shows him struck by the ease of his forty-minute flight over desert where the children of Israel had wandered for forty years. Baghdad, faded, decayed, mean and squalid retained its Oriental mystery and allure. Smith remembered the *Arabian Nights,* Aladdin and Sinbad. Flying over Kut-el-Amara reminded him of its heroic defence by the British, and their valiant surrender. He was overhead another imperial military battlefield reminiscent of Lucknow, Delhi, Khartoum, Ladysmith and Mafeking. Over territory that was once fair Babylonia, Smith mused on the ravages and misrule of foreigners which had turned a land of milk and honey, pomp and luxury, into a void of marsh and waste.[62] Lord Thomson reflected only on the ease of observing ancient history within a week of leaving any European capital: his October 1924 RAF flight from Cairo into the mandated territory of Transjordania showed the ruins of Babylon, Ninevah, Palmyra and Ur.[63]

In the 1930s, the *Sphere* told its readers that at Gaza *Imperial* passengers were staying where Samson had once removed the city gates. The flight from Jerusalem to Baghdad was said to follow the Jericho road and gave a view of the little village of Bethany, the Apostles' Fountain and the Good Samaritan's inn. Antiquarianism, orientalism and piety were served well. In 1934, a series of articles published at weekly intervals in *The Times* pointed out how flying made it possible to see strange, sacred, mythic lands. The air route was not just utilitarian; it was also an education. Second best to having schoolmasters point out sights from flying classrooms, the newspaper's photographer (using 'British apparatus in a British aircraft') produced pictures of the Gulf of Corinth, ancient Athens, Bethlehem and Jerusalem. A contemporary tome about international affairs noted that *Imperial* aircraft roared along a route where once Alexander the Great had marched and sailed.[64]

Natural landscapes were, of course, a component of imperial

geography. The 1934 *Times* series opened with a double-page aerial photograph of Jabal-i-Mehdi, the massive geological outcrop along the Baluchistan coast near Gwadar. The view of the distinctive spires and pinnacles that made up the craggy formation known in English as the Ass's Ears or the Cathedral Rocks was unsullied in nature. On paper the mountains were surmounted by a half-page advertisement for Shell Oil in which prominence was given to a sidelong view along the wing of an *Imperial* aircraft. The commercialism detracted from what was widely understood to be 'the romance of flight'. People could regard scenery differently, of course: the Cathedral Rocks gave one airline steward 'the creeps'.[65]

In an article marking the start of *Imperial*'s first flying boat through-service from Southampton to Darwin in 1938, the *Sphere* published an aerial photograph of the great ruin of Agar Kuf near Baghdad.[66] Seeing this relic (and other antiquarian sites) with his naked eye might be expected to have pleased the twenty-five-year-old classical scholar on the plane, Enoch Powell. Brilliant but flawed, as an elder British politician he would be denounced for his bigoted attitudes to foreigners. One of only two through-passengers, he was on his way to taking up his post as Professor of Greek at the University of Sydney. His first personal flight was the first of four England–Australia flights he would make with *Imperial*. On this occasion Powell seemed neither interested in nor amused by Greece. His detachment would have been incomprehensible to the three-man airline crew and the three women passengers and young boy who were photographed with him as he showed off the Acropolis.

Powell's gruelling, sixteen-day journey (including two days grounded for mechanical repairs and transfer from flying boat to landplane at Singapore) was less an encounter with classical civilisation than a 'revelation' about the extent and majesty of contemporary Empire. His biographers note the stirring 'lesson in imperial geography' that the flight taught him. Landing four or five times a day, nearly always on British territory, was an experience that he never forgot. The early morning starts, and the insect-infested table cloth in the refreshment hut in Timor, paled beside his growing awareness of the ubiquity of British power: 'I saw; I felt; I marvelled.'[67]

Settlements *en route* to India were a part of a more-or-less tamed landscape, and spending time in them was part of the sociability of flying to and from the East. If Dower's impressions of Baghdad in 1932 are anything to go by, *Imperial* had a job on its hands redeeming something from the 'nightmare place', its 'ghastly squalor', 'the evil crowd of sick villains on the streets' and the detritus of 'a civilisation of discarded petrol tins'. The 1933 over-Everest team had been warned: the palaces, gardens and courtiers of the Baghdad immortalised in the

'Arabian Nights' had been replaced by 'hovels feeding on the dole of past greatness'.[68] *Imperial*'s passengers stayed overnight at the Hotel Maude and had the use of the town's three main clubs. In descending order of status, these were the 'British', the 'Railway' and the 'Alwiyah'. The first was exclusively male and was used mainly by city gents. The other two had a mixed membership and offered facilities for swimming, tennis, squash, polo, wild boar spearing and game shooting. At Basra, passengers stayed near the airfield at the Iraq Railway Rest House.[69] Club and railway were known and trusted Empire institutions. Like seaports and railway stations before them, airfields were the new imperial bridgeheads and sanctuaries. But flying could also bring unwanted novelty to clubs started by expatriate residents 'to ward off dullness': while on her 1939 flight, Eileen Bigland felt the disapproval of her socialising with the Indian editor of the *Karachi Gazette*.[70]

Flying east from Basra, conditions deteriorated. Where luxury was impossible, it was good enough that there was attentiveness and effort. Even modest ground facilities in remote and hostile environments signalled imperial achievement. Security took precedence over comfort in some places. The stone and cement rest house at Sharjah was no ordinary building. Rather, it was a beau geste fort with watchtowers, parapets, iron-spike palings and barbed wire. Three-dozen guards, retainers of the local Sheik, manned it. The defence was deemed necessary as the landing ground was in one of the 'wild and uncivilised parts' that inevitably would be crossed by any 'vast civilian air organisation that knit the Empire into one impregnable whole'. Where neighbouring tribes were hostile the airline had to be defended against raids.[71]

The practical implications of such isolation were that stores would have to be taken to landing grounds far from road or railway. Repairs would have to be undertaken many miles from factory or depot. At Sharjah, drinking water was delivered on donkey back in water skins. Petrol was offloaded from dhows and forwarded by camel. The Spartan conditions could be interpreted as a sign of vulnerability, but when Sir Montagu de Pomeroy Webb passed through in 1932, he declared the rest house 'a triumph for British enterprise and organization'. In the air and on the ground, passengers doubtless appreciated all signs of control and safety, while nevertheless deriving some thrill from being on the 'Pirate Coast'. Even the indignity and irritation of a forced desert landing beside an uninhabited, roofless fort could be made into an amusing adventure and could galvanise camaraderie. In just such a situation, one *Imperial* pilot sent his passengers out to collect dry dung in their hats so that they could keep warm during the cold nights.[72]

Imperial preferred formal arrangements for its passengers, of course. Occasional informal hospitality could get entrenched. When the first commercial flight landed at one refuelling stop in the Middle East, the British Resident apparently invited the crew and passengers to breakfast. He continued to do so for a year, the flight captain assuming that this entertainment had been organised by the airline.[73]

Personal 'violation' during travel away from base did present in ways not meriting elaborate fortifications. In March 1933 the Political Agent at Bahrain reported to the Resident, Lt-Col T. C. Fowle, that one of *Imperial*'s women passengers had gone into Sharjah town wearing casual clothing described as 'beach pyjamas'. He did not want visitors to affront local people and destabilise *Imperial*'s toehold. Enigmatically, he also did not wish to 'spoil' the Sheik. Fowle, whose conviction was that 'he represented the British civilising mission in a still semi-barbaric land', lodged a complaint with *Imperial*'s Rest House Superintendent and with the airline management in London: Sharjah's people were unused to strangers, he wrote, especially women wandering their bazaars. He did not want passengers courting harm. Fowle warned the Sheik that he would be responsible if any passengers were 'insulted or molested by some bad character, or Bedouin from the desert'. The concern has also been interpreted as managing disruption of an ordered male world by an increasing number of people travelling independently beyond conventional British political and diplomatic controls.[74]

Conditions along the India air route became harsher further east. The Gwadar air station, eight miles from the town at the end of a sandy pathway, was utilitarian only. Clustered around it were an electricity generator, a wireless room, a small rest house, a petrol store and a floodlight shed. The handful of expatriate staff at this outpost of Empire included three Indian meteorologists, the British agent (generally a retired Indian Army officer) and the representative of the Anglo-Persian Oil Company. The Superintendent at Gwadar regarded the group as 'a new class of pioneer'. They, if not their short-stay Empire visitors, ate mostly local fish, milk, eggs and scraggy chicken. Tinned foods and a basket of fresh fruit and vegetables were flown in weekly from Karachi.

Passengers and crew were not the only consumers of stores at Gwadar: a 'native' easily unfastened the shed padlock without a key (and required no obvious payoff) when a hungry C. W. A. Scott wanted food during his record solo flight to Australia in 1932. *Imperial Airways* never billed Scott against the IOU note he left for a tin of sausages. Drinking water was obtained fortnightly from the British India Steam Navigation Company boats. The local water supply, of dubious hygiene, was only used for bathing and cooking. Recreation included tennis, golf and bird shooting. An employee's gunshot once injured a local person accidentally.[75]

Flights to and from the East in the 1930s increased in speed and frequency, and refuelling and (off)loading and overnight stops changed in response to shifting *Imperial Airways* flight plans. Monthly reports filed by Britain's Political Agent at Kuwait toward the end of the decade show frequent unscheduled bypassing of the town. Modernisation did not wholly wreck the romance of flight however, as the diary-like account of a staff writer on *Flight* magazine shows. Printed in two consecutive issues in September 1938, it relates to the newly inaugurated flying boat service. A route map, photographs and anecdotes accompanied the text. A keen eye spotted the platform sign at Waterloo station that signified the departure time and platform for the 'Empire Air Service'. The Southampton boat train had a competitor: the Southampton flying boat train.

A flight to the East could itself be an education. A pamphlet issued by *Imperial Airways* acted as mute tour guide. 'After Syria we skirt the Hauran, formerly the land of Bashan, peopled by a race of giant men, and now the mountain stronghold of the Jebel Druses.' The Gurkha at Allahabad guarded the aircraft with his *kukri*. At Dubai, the same task fell to an Arab who fondled an old automatic gun. The writer made no comment on any security arrangements which the Political Agent at Bahrain had sought when, a year earlier, he arranged a meeting with the two relevant Sheiks to negotiate the safety of *Imperial* passengers shuttling between the landing area in Dubai and the rest house in Sharjah.[76]

The discovery of curious places was bread and butter to the prolific American travel writer, E. Alexander Powell. In the 1930s he flew as guest of and companion to the Chairman of a large Chicago industrial company. One stretch of their long journey by air involved taking *Imperial Airways* westward from New Delhi to the Persian Gulf. There, one of the carrier's 'primitive' rest houses sheltered them from a sand storm which obliterated the sight of their aircraft from more than thirty feet away. The passengers ate, slept and played bridge on tins of petrol in the supply dump. They flew onward on *Hadrian* to Sharjah, where *Imperial* had built a beau geste fort similar to the one at Rutbah Wells, garrisoned by a detachment of Arab irregulars, encircled by barbed wire.[77] It was during an overnight stop at Sharjah that Hudson Fysh expressed relief at being on the other side of wire entanglements from '200 of the most magnificent Arab ruffians it is possible to imagine'.[78]

Perspectives

Lord Lothian (fifty-six), the aristocratic bachelor politician and diplomat (whose imperial career spanned five years' government service in South Africa and one as Under-Secretary of State for India), spent several days as an *Imperial Airways* passenger while flying from London to Sydney and back in 1938. Immediately before taking up his post as British Ambassador to Washington, it was flight that enabled him to keep his prior commitment to travel on Rhodes Trust business and also chair a British delegation to a conference at Sydney University on Commonwealth Relations. His 35,000 mile journey, he wrote, gave him a new perspective on the past and present. For about twelve hours, flying – and his Oxford education – enabled him to 'look in' on four of the 'great civilisations' of ancient and modern Europe, namely Greece, Rome, France and England. The next day he flew over the civilisations of Crete, Egypt, ancient Palestine, and the empires of the Assyrians, Medes and Babylonians. It took just a day to pass Delhi, Agra, Benares and Allahabad, the capitals of emperors, kings and some of the most enduring religions. In compressed time, Lord Lothian reached Australia – 'the most mechanistic civilisation' in the modern world. Flying had laid bare the panorama of history, from stone age to skyscraper, in less than ten days.[79]

Grace Crile also remarked on how Empire passengers were carried from 'the greatest concourse of animal life' across the seats of ancient powers and civilisations in Alexandria, Crete, Athens and Rome, to 'the great concourse' of contemporary power and culture in Paris and London. In reverse, another contemporary writer noted how a week's journey to Rhodesia traced the history of civilisation via Rome, Greece, Crete and Egypt to the spot where humans first emerged from the apes.[80]

The first moral of his flight, Lord Lothian felt, derived from the debris of cities, irrigation works and temples. Once filled with people, they now lay empty and dead: 'thus passed the glory of the world!' Might modern technology save the British Empire from the same fate? The answer was unclear at the time. More immediately apparent was disagreement about flying sensations and sights. In 1939 Bigland noted that most people praised the speed of air travel but condemned it because 'you could never see anything'. She, on the other hand, regretted the all too brief view it gave of the countries passed over.[81]

A vertical perspective also disappointed an *Imperial* passenger flying from London to Salisbury who regretted his first sight of the pyramids. From a height 'they looked like a child's bricks'.[82] On his six-day south- and north-bound *Imperial* flights across the length of Africa in 1933, the lecturer-writer Maurice Samuel found the Nile 'fragile and provisional and inconsequential' from the air. Looking down on Rhodes' open air mausoleum he saw 'a tiny, impertinent toy, a pygmy gesture'.[83] The 'pageant of nature' and 'magic carpet' which one passenger enthused about in 1936 did include reference to Mount Kilimanjaro as 'an enormous Christmas pudding covered with white icing'. Rice fields at Bangkok looked like billiard tables side by side, and the Irrawaddy River at Rangoon appeared to be a double cracker. This was indeed 'seeing the world through different eyes'.[84] The time taken to cross vast landscapes even at speed put territorial size into perspective: the Syrian Desert (the Transjordania mandate) is larger than the combined area of England, Scotland, Wales and Ireland; the Libyan Desert is the size of India.

Like airline passengers, private pilots were also ambivalent about the sights they saw. Some resented the experience because it sullied imagination and flattened majesty. Francis Chichester's 'romantic notions' about Egypt were quickly dispelled by the sight of villages and towns with thousands of mud shacks jammed together in the 'boring' Nile Delta.[85] Alan Cobham found Victoria Falls very disappointing from the air, resembling 'a futile little river running into a crack in the earth'. But he too contradicted himself when reporting an 'inspiring spectacle' in a telegraph from Bulawayo.[86] The Duchess of Bedford found the Falls unimpressive from above, and wanted to see them from below. With a better view, her pilot thought them 'easily the most impressive' aerial sight he had ever seen.[87] The Holy Land could be very disappointing. On her 1928 flight to India, the Duchess remarked that 'people who cherish the allegory of the Garden of Eden had better not visit this Biblical Paradise'. Dower was sharper still, dismissing it as 'rather cheap blasphemy' and 'a pathetic fifth-rate sort of affair'.[88] This corrective to the standard romanticism was not the kind of publicity *Imperial* seized on to promote flying over the Promised Land.

Perspective was about history, angle of vision and normalisation. One passenger found that the *Challenger* flying boat's inaugural 13,000 mile, ten-and-a-half-day England–Australia flight in 1938 entailed 'a compete lack of adventure'. A press report added that 'the flight was almost dull' and lamented that high adventure had become efficient routine: 'it is on reality that the imagination feeds, and the air passenger's detachment from earth-bound things severs him from reality'. Aloft and insulated, passengers' imaginations refused to digest the diversity of sights, smells and landscapes: 'the living map, unrolling swiftly thousands of feet below him, becomes, by the very speed of his transit, little more stimulating to his curiosity than the wall maps of his earlier school room. Man's mastery of the air has taken the romance out of the phrase "to the ends of the earth". We adjust our ideas and welcome the globe's shrinkage.'[89]

Managing uneventful journeys might have become a concern for *Imperial Airways* publicists toward the end of the 1930s had war not been distracting. It was fine for a passenger's diary of a flight home to Australia and back in the space of three weeks in 1938 to record that 'old ideas of distance have crumpled up' and that one could be over three continents (Europe, Africa and Asia) on one day. But the 3 a.m. starts were hardly an attraction when the cafés in Athens were still full of men drinking and talking. And it was poor publicity to report that beyond Basra there were only four passengers on the *Imperial* flying boat. Worse, though, was the comment that their curiosity had faded by then: 'as the boat whizzed over the world . . . one of them was playing patience and two chess'.[90] Fellow passengers could be distractions themselves, of course. Bigland also contradicted herself. On her 1939 flight east from Baghdad, she conceded that the 'mixed bag of passengers' once again 'destroyed my interest in the landscape'.[91]

Mostly, the rapid cultural juxtapositions and fleeting contacts which flights across the Empire made possible did not alter the way the British world was read and related. Few air passengers seized the opportunity to embrace new modes of thought and behaviour. Few questioned whether flying over a country meant losing even more contact with local people than during overland rail journeys when visitors travelled through rather than among communities. Writings about Empire air journeys were in the tradition of Victorian armchair travel literature in which readers shared the experiences of the writer vicariously. The writing itself presented the world beyond in a matter of fact way, seen only through Western eyes. Whether by air or overland, foreign travel was to, and with, a set of familiar British overseas contacts who enjoyed a carefully nurtured insularity from indigenous people who continued to be presented mostly in caricature.[92]

One overland traveller did pause to consider how flight was minimising contact with the 'true' Africa. Capt. Reynardson, who, with his family, drove from Cape Town to Mombasa in 1933, reflected on the three days which the 'air mail' passengers had taken to get from Cape Town to Mbeya. His party had taken three months over the same journey. He broke the mould by tracking the airway not the railway for some stretches, and by using airfield – not railway – rest stops, hotels and post offices as places to buy supplies, and to collect and dispatch mail. It was easy to feel inferior about one's vehicle after seeing *Imperial*'s 'immense and graceful dragonfly' landing at Mbeya, and it was easy to feel like a slovenly tramp among the smartly dressed *Imperial Airways* passengers dining in the local hotel, but slow was not necessarily worse. Indeed, Reynardson thought that air travellers missed out: they were part of the new Africa, 'but what had those passengers in their "travelling costumes" and natty tweeds seen of the Old?'[93]

The literature of flight persisted in accentuating the 'otherness' of non-British people overseas, implying superiority of those who had fallen under the 'civilising spell' of the British Empire. The reality and the textuality of Empire civil aviation were complicit in co-constructing categories of primitive and modern, and thereby helped to constitute Empire. Anecdotes about Africans and aircraft make good examples from the late 1930s. After seeing an *Imperial Airways* aeroplane, an elderly African 'chieftain' sought an audience with the Company to ask if he could acquire an egg from the great bird for propagation. Bird imagery was central to people living simple lives close to nature. When Africans saw flying boats for the first time in 1937 at stops on the Tanganyika and Mozambique coast, they either named the aircraft generically as 'the great white birds' (in Swahili) or, in pidgin, as 'imperiali'. The motor launches that serviced the flying boats were called 'sons of the birds that fly'. The parallel with nature served imagination only up to a point. Africans were reportedly amazed that every time a launch set out a flying boat would emerge. The magical coincidence earned for *Imperial* staff the nickname 'masters of the birds'.[94] The passivity of airline employees in this unoriginal but supposedly spontaneous reinvention of avianisms is questionable in light of their widespread and long-standing usage. By 1937, even English-speakers with the poorest ear for African languages might have been responsible for spreading the various appellations.

There were exceptions to writing that mostly failed to pose challenging and unsettling questions about flying. Maurice Samuel and one of his South African admirers fretted about it reducing Africa to a relief map, diminishing it, and disorienting and distancing Europeans and Africans.[95] In 1936, after seeing tall, naked, 'jet-black' warriors dancing and chanting at Malakal, one passenger-writer wondered if they would be happier or better when 'civilisation' taught them shame and sold them short trousers.[96] The American, Grace Crile, also paused to puzzle, despite having her share of stereotypes. Like Sir Alan Cobham, A/Cmdre Samson and Lady Heath (and, surely, C. G. Grey if he had flown), she made the customary judgements. Yet, while reporting Africans to be dirty and 'distasteful', Crile did ponder the superiority that she felt. 'We send missionaries out to Christianise them, yet, left to themselves, they have codes of morals and justice which they, as tribes, obey and revere', she observed. 'Does Civilised Man do more?', she wondered.[97]

How did aviation stand in this civilising mission? Was it uplifting in only the technical sense, and then only for some people? The answer given by a talented thirty-five-year-old British Africanist reflected poorly on the White people she met during her lengthy Rhodes fellowship study tour of East Africa in 1929/30 (excluding, presumably, her brother-in-law, the Provincial Commissioner). Margery Perham's view of the contrasts between Africa and Europe that were evident at airfields in the African wilds was that they would be less superficial and more startling if only the people who landed did not turn out less godlike than their descent from the clouds promised.[90]

Notes

1 A journalist reporting his first flight, escorting Cobham from Pretoria to Johannesburg. *Rand Daily Mail* (Johannesburg) (8 February 1926).
2 *Imperial Airways Gazette* (April 1930; February 1932); H. E. Brittain, *By Air* (London, 1933); H. Yea, 'Air mail to the Cape' (unpublished paper, n. d.).
3 *Imperial Airways Gazette* (June 1933); J. Charmley, *Lord Lloyd and the Decline of the British Empire* (London, 1987), pp. 184, 209, 213.
4 L. Farrant, *The Legendary Grogan* (London, 1981); E. Paice, *Lost Lion of Empire: the Life of Cape-to-Cairo Grogan* (London, 2001).
5 *Imperial Airways Gazette* (February 1932); *The Times* (2 April 1934).
6 M. de P. Webb, 'India to England in five days: a marvel of modern transport', *Asiatic Review*, 28 (1932), 123–8; D. Kincaid, *British Social Life in India, 1608–1937* (London, 1938), pp. 274–5.
7 *Field* (6 July 1935), p. 47.
8 *Time* (27 June 1938), pp. 30–1.
9 *The Times* (11 November 1936), p. 23.
10 *The Times* (20, 22 July 1932); (11 November 1936), p. 23; A. Frater, *Beyond the Blue Horizon: on the Track of Imperial Airways* (Harmondsworth, 1986), p. 80; *Sphere* (13 June 1931), p. 470.
11 *Popular Flying* (August 1936), pp. 237–40, 264.
12 *Manchester Guardian* (21 January 1932), p. 8.
13 *Imperial Airways Gazette* (January 1934); *The Times* (12 October 1933), p. 9; F. M. C. Stokes, 'Flying back through history', *Blackwood's Magazine* (1936), 684–91.
14 *The Times* (20 July 1932), p. 8.
15 *Evening Post* (Wellington) (8 December 1934), p. 8.
16 *Imperial Airways Gazette* (January 1934).
17 Kenworthy, *Sailors*, p. 227; *Sphere* (21 December 1935), p. 466.
18 PDPG (May, June 1934; August, September 1936); *Canberra Times* (31 August, 3 September 1936); *Sydney Morning Herald* (14 September 1936), p. 10.
19 BL, OIOC, R/15/2/514(13/5).
20 *Flight* (26 January 1933).
21 *Mercury* (Hobart) (2 September 1932), p. 7.
22 H. Klein, *Winged Courier* (Cape Town, 1955); Coster, *Corsairville*; Sims, *Adventurous*, p. 120.
23 G. Pett, *Beyond the Blue Horizon: Transcripts and Memories of Imperial Airways in Africa* (1995); http://geoffreysbox.wordpress.com (accessed December 2010).
24 W. Hichens, 'Bush-lore for the airman', *Nineteenth Century* (1930), 548–55; Russell, *Flying*.
25 *Flight* (28 October 1937), p. 427; *Listener* (22 September 1938), pp. 598–9; Sims, *Adventurous*, p. 114; P. J. Davis, *East African: an Airline Story* (Egham, 1984), pp. 15–16.
26 E. Huxley, *Out in the Midday Sun* (London, 1985); D. C. T. Bennett, *Pathfinder: a War Autobiography* (London, 1958), p. 64.
27 *Imperial Airways Staff News* (8 July 1938).
28 Hudson and Pettifer, *Diamonds*, pp. 67–8, 70.
29 J. A. Mollison, *Playboy of the Air* (London, 1937), p. 266; Luff, *Mollison*, p. 289.
30 K. Hudson, *Air Travel: a Social History* (Bath, 1972), p. 103.
31 Pilkington Archives (St Helens), Leaves from W. H. Pilkington's diary of his air trip to and from Australia; Hudson and Pettifer, *Diamonds*, pp. 64–8.
32 *Journal of the Royal Aeronautical Society*, 32 (1928), 4.
33 *Aeroplane* (6 July 1932), p. 8.

34 *The Times* (19 August 1931), p. 6.
35 Morris, *Farewell*, pp. 359–60.
36 W. J. Makin, *Swinging the Equator* (London, 1935), p. 189.
37 *Imperial Airways Gazette* (August 1936), p. 4.
38 Makin, *Swinging*, p. 235; Crile, *Skyways*, p. 210; Royal Air Force Museum (RAFM) (Hendon), AC 76/49.
39 Sudan Archive (Durham University) (SAD), 707/15/16–17; Crile, *Skyways*, pp. 34, 40.
40 *Imperial Airways Gazette* (October 1932); A. Warwick, 'The great trans-African airway', *Windsor Magazine* (1932), 427–39; Crile, *Skyways*; Hudson, *Air Travel*, p. 97; R. U. Light, *Focus on Africa* (New York, 1941), p. 202.
41 R. Brenard, 'The romance of the air mail to East and South Africa', *Journal of the Royal African Society*, 38 (1939), 47–64.
42 M. A. U. Heathcote, 'By air mail to Kenya', *Geographical Journal*, 79 (1932), 502–6.
43 Crile, *Skyways*.
44 *Africana Notes and News*, 26 (1985), 261–2; Cobbold, *Kenya*.
45 *Imperial Airways Staff News* (12 May 1936).
46 A. R. Cooper, 'Flying to India', *Geographical Magazine*, 1 (1935), 141–50, 173–83; A. R. Cooper, 'Flying to India', *Geographical Magazine*, 2 (1935), 67–78; A. R. Cooper, 'Flying to India', *Geographical Magazine*, 3 (1936), 218–30.
47 H. Butcher, 'Cape to London airway', *Chambers's Journal*, (1932), 56.
48 Coster, *Corsairville*; SAD, 769/9/9.
49 W. Thesiger, *The Life of My Choice* (London, 1987).
50 South African Air Force Museum (Ysterplaat) (SAAFMY): Miller Papers, R43-2.
51 South African Library (Cape Town) (SAL): MSB 750.
52 N. Nelson, *The Shepheard's Hotel* (Bath, 1960); Cambridge University (Manuscripts Reading Room) (CU): Templewood V (8).
53 Trzebinski, *The Lives of Beryl Markham*, p. 197; *Field* (13 August 1932), p. 256.
54 Huxley, *Out*, pp. 90–1.
55 F. A. Smith, *White Roots in Africa* (London, 1997), p. 54.
56 E. Huxley, *Nellie: Letters from Africa* (London, 1980), pp. 92–3; 109–11.
57 H. J. M. Camac, *From India to England by Air* (New York: privately published, 1929).
58 F. K. Ward, 'Linking the Empire by air: to India in seven days', *Empire Review*, 50 (1929), 419–27; A. Eggar, *The Indian Air Mail from Croydon to Karachi* (privately published, 1929); *The Times* (19 June 1929).
59 *Daily Express* (11, 12, 13, 17 September 1929).
60 *New Statesman* (14 June 1930), pp. 298–300; R. Byron, *An Essay on India* (London, 1930); P. Fussell, *Abroad: British Literary Travelling between the Wars* (Oxford, 1980), p. 89; E. Thompson, *You Have Lived Through All This* (London, 1939), p. 85.
61 C. Roth, *The Sassoon Dynasty* (London, 1941); Nicolson (ed.), *Harold*; P. Stansky, *The Sassoons* (New Haven, CT, 2009).
62 Smith, *14,000 Miles*.
63 C. B. Thomson, *Air Facts and Problems* (London, 1927), p. 136.
64 *Sphere* (13 May 1933; 17 March 1934; 22 January 1938); *The Times* (8, 15, 22, 29 March 1934); S. King-Hall, *Our Own Times, 1913–39* (London, 1935), p. 601.
65 Bigland, *Awakening*, p. 138.
66 *Sphere* (22 January 1938), p. 144.
67 R. Shepherd, *Enoch Powell* (London, 1996), p. 33; S. Heffer, *Like the Roman: the Life of Enoch Powell* (London, 1998), pp. 36–7.
68 Dower, *Amateur*, pp. 165–7; Fellowes, et al., *First over Everest*, p. 127.
69 Frater, *Beyond*, pp. 87, 94; H. C. Sinderson, *Ten Thousand and One Nights* (London, 1973), pp. 72–3.
70 Bigland, *Awakening*, p. 139.
71 R. E. H. Allen, 'Ground transport for an air organisation', *Journal of the Royal Aeronautical Society*, 32 (1928), 596–616.
72 NA, AIR 5/1217; *Sphere* (4 March 1933), p. 292; M. de P. Webb, 'Links in the Imperial chain', *Asiatic Review*, 29 (1933), 693–7; Frater, *Beyond*, p. 80.

73 Morley, *Colonial*. This recollection by the twenty-one-year-old on his way to Aden is of the Residency on the Island of Kamaran in the Red Sea. *Imperial* did not operate any scheduled services to or through Kamaran, but the stop may have been on charter flights to Aden.
74 P. Tuson, *Playing the Game: the Story of Western Women in Arabia* (London, 2003), pp. 208–11.
75 Scott, *Scott's Book*, p. 233; A. E. Thompson, 'An air mail outpost', *Air and Airways*, 9 (1932), 233–4; NA, FO 371, 16834 (E2373/11/91).
76 King, 'Empire'; PDPG (September 1937).
77 E. A. Powell, *Adventure Road* (London, 1955), p. 284.
78 *Courier-Mail* (Brisbane) (6 September 1933), p. 15.
79 J. R. M. Butler, *Lord Lothian (Philip Kerr), 1882–1940* (London, 1960), p. 225; *Observer* (27 November 1938), p. 12.
80 Crile, *Skyways*; Stokes, 'Flying'.
81 Bigland, *Awakening*, p. 129.
82 *Imperial Airways Gazette* (January 1934), p. 2.
83 M. Samuel, 'Flight across Africa', *Harper's Magazine*, 169 (1934), pp. 32, 36.
84 *Observer* (7 June 1936), p. 28.
85 Chichester, *Solo*, p. 92.
86 *Rand Daily Mail* (Johannesburg) (4 February 1926); *Star* (Johannesburg) (1, 2, 4 February 1926).
87 Gore, *Mary*, pp. 277, 278.
88 Gore, *Mary*, p. 205; Dower, *Amateur*, pp. 144–5.
89 *Observer* (10 July 1938), p. 16.
90 *Manchester Guardian* (5 August 1938), p. 8.
91 Bigland, *Awakening*, pp. 140, 143.
92 H. J. Field, *Toward a Programme of Imperial Life: the British Empire at the Turn of the Century* (Oxford, 1982).
93 Reynardson, *High*, p. 156.
94 Pudney, *Seven*; Pett, *Beyond*; *Saturday Review* (14 August 1937), p. 107.
95 Samuel, 'Flight', pp. 30–9; E. Lewis, 'Flight over Africa', *African Observer* (1935), pp. 55–8.
96 Stokes, 'Flying', p. 688.
97 Crile, *Skyways*, p. 207.
98 M. Perham, *East African Journey* (London, 1976), p. 227.

7

Personifying Empire

The image of *Imperial Airways* as an organisation, and its iconic status in the Empire, hinged partly on its perceived efficiency and reliability, and partly on the impression created by its senior management. In one of many such public displays, *Imperial*'s Chairman, Sir Eric Geddes, articulated a glowing, benevolent alignment between aviation and Empire in a speech to the Imperial Press Conference at the Carlton Hotel in 1930. Speedy air transport, he said, was the most important consideration to those who loved the Empire and wished its strengths to develop.[1]

Further away from the levers of power (but closer to the controls of everyday flight, where actions spoke louder than words), the flying crew and ground staff personified the values of their employer and the Empire. It was a load shared among a relatively small number of employees. At its inception in 1924 the airline employed 260 people, nineteen of whom were pilots. By 1928, when air services were still wholly within Europe, the tally was 368 in summer and 288 in winter. By 1933, when imperial air services had started operating, the number of people on the airline's payroll had increased to 1,200. There were approximately 300 ground personnel on the African route.[2] In 1936, more than thirty *Imperial* pilots worked the section between the Mediterranean and Karachi or Kisumu. Almost twenty flew the Karachi–Singapore and Penang–Hong Kong routes. Ten pilots were stationed on the Kisumu–Johannesburg route. On the eve of the upgraded flying boat service linked to the new Empire Air Mail Scheme (EAMS), the airline was reported to be strengthening its complement of sixty pilots by taking more into training.[3]

Imperial's employment roll was geared to its own operations first and foremost and not to the projection of British imperialism. But employee numbers and reward were certainly one face of Empire, and one that could grimace. Toward the end of 1936 *Imperial*'s senior engineer alerted airline managers to problems of recruiting and retaining overseas crew. Only months before the start of the EAMS, which was estimated to require 200 more contract staff than the 187 then employed, there were only twenty men in training for the Empire services. Hours of work were one cause of dissatisfaction. Another was the slow rate of promotion to what was termed 'married strength', presumably a closed-sized group of employees whose wives could live with them overseas. The level of the 10s weekly 'separation allowance' for married men stationed overseas alone was another point of contention. *Imperial*'s wages were said to be less than the skilled British aeronautical wage of £5 weekly, and the overseas pay was judged insufficient to compensate for the risks and discomfort. The latter was physical and social: employees abroad complained that they were 'excluded from white circles and would not enter native ones', even if it were possible.[4]

Cockpit crew

The new technology of flying did not corrupt the rigidly stratified social organisation so characteristic of 1930s Britain at home and abroad. Aircraft captains were at the top of the pile as regards responsibility, status, social class and earnings. Navigators, flight stewards and ground crew ranked more humbly.[5] Speed alone did not warp the age-old distinctions that applied to the Empire's maritime and overland transport arms. Indeed, Empire aviation traced the tiers, ceilings and colours of transport work in colonial shipping and railways.

Not unlike their equivalents at sea, Britain's Empire airline pilots were vested with status and, mystique. Their small number meant that individuals even had a public following. Their photographs appeared in magazines and books, and, for a time, disembarking passengers were handed autographed photographs of their captain as a flight memento. In 1932/33, *Air and Airways* ran a series of feature articles about 'Pilots of the airlines', several of whom worked for *Imperial*. Pen-portraits accentuated their modesty and capability, and spoke of the camaraderie among a new breed of lone, self-sufficient pioneers. The men up-front were not of this grubby world; they were 'the salt of the sky'.[6] As late as 1938, six years into Empire commercial flying, the obituary of an *Imperial Airways* pilot (but not his four crew) could appear in *The Times*. Capt. Ernest Attwood, who died in the *Calpurnia* flying boat accident on Lake Habbaniyah in Iraq, had been with the airline since 1926. He was remembered as experienced, highly qualified, careful, conscientious, cheery and companionable. The thirty-nine-year-old, born in Birmingham and schooled in Aberystwyth, left a widow.[7]

Something of the reverence enjoyed by Empire-line pilots was a matter of costume and style. As the manager of the Aerodrome Hotel at Croydon observed, all English pilots looked like merchant navy skippers. It helped that from early 1931 officers on *Imperial*'s 'mercantile air service' wore half-inch-wide gold bands on their sleeves, aping officers at sea.[8] On a scale rather less than his rhetoric would have suggested, Brancker's 1925 wish was beginning to come true: a nation of seamen was becoming a nation of airmen.[9]

The pilot whom *Imperial Airways* liked to promote as stereotypical was Capt. O. P. Jones. A biographical sketch described the thirty-five-year-old as bearded and fit. He was made unmistakably masculine. He lived a clean and healthy life: he was teetotal, and liked to walk his dogs. His Bentley sports car (which he drove up to his waiting aircraft) told of a solid citizen and a man of taste. Put him in his work uniform, and he was easy to portray as 'the true ship's captain of the Merchant Service of the future – that of the air'. His symbols of authority were a cap and a spotless, smart, dark blue uniform with three golden sleeve rings. He liked to wear white gloves while flying.[10] In his impeccable dress and bearing he resembled Capt. Kettle, the seafaring hero of Cutcliffe Hyne's boys' stories of the 1890s. A contemporary sketch writer made the association without hesitation: Jones carried himself with 'the spruce, ship-shape manner of the little sea-captain of fiction'. Whether the writer would have wished him to possess all the attributes of the archetypal 'romantic rogue' who 'licked the world into shape for England' is a moot point. Certainly it was desirable that he be seen as fearless, independent, reliable and proud. If only unintentionally and subliminally, a latter-day Kettle was also 'the Englishman re-energised, ferocious yet clean, savage but white'.[11]

Promoters of Empire aviation, knowledgeable commentators and enthusiastic air travellers used maritime metaphors avidly. Inadvertently or not, aviation was also used to catch glimpses of a mythical past, or indeed to reconstruct history. As late as 1937, the American Grace Crile wrote of her *Imperial Airways* flight to Africa that the captain and crew continued a long and honourable tradition. The 'dauntless young navigators of the air are worthy representatives of the heroic explorers and adventurers of all the ages', she claimed. Like their predecessors, they were also 'adding page upon page to the romantic history of all time'.[12] Arguably, coachmen who drove teams of horses across rough, unmarked tracks in the eighteenth century deserved greater accolades; sea captains in their wooden world most certainly did. Nevertheless, the notion died hard that the air was a particularly thrilling, mysterious medium, and that the world which was wide of Greenwich by more than a few degrees longitude was a dangerous place. British air pilots, a new class of peacetime officer-heroes, made the new world safe.

Mostly, that is how it appeared; there could be hiccups. In 1936, for instance, there were allegations that Italians had meted out 'brutal treatment' to Capt. Walter Rogers and his crew of three aboard an *Imperial* aircraft delivery flight to Cairo for service in the Near East. Landing at Mesylam, Libya, the men were placed under armed guard 'in an inadequate tent in the blazing sun with armed native guards outside'. The four were allegedly paraded as objects of derision at bayonet point in front of Italian troops. When their seized papers were found to be in order, the Italians claimed that the plane had been flying on a prohibited track.[13]

Imperial's pilots (sixty flying boat captains and first officers on the Empire routes in 1936) were carefully selected and expensively trained (in aeronautics if not in political negotiation). They were paid well, earning up to £1,300 annually in 1939.[14] They were certainly not mere engine drivers. 'The pilot of an *Imperial Airways* machine is in every sense a flying ambassador', wrote Sir Harry Brittain in his 1933 aviation book for youngsters. They were 'ambassadors of England' said one pilot, even though *Imperial* told its pilots to be hosts and did not mention ambassadors. Sir Harry did not dwell on these roles, but moved on smartly to say that the airline took 'as much pride in its personnel as in its machines'. Elaborating, he noted merely that they were all British. This was just as well for, as a carrier of royal mails, *Imperial* pilots were an unofficial agent for the Crown. Re-creating the spirit of Empire, Jan Morris records that if an *Imperial* captain made a forced landing in imperial territory he was authorised to stop any passing train and oblige it to take the mailbags.[15] Apparently, packages were more important than passengers, for the pilot's semi-regal authority did not extend to his majesty's human subjects. There were even occasions when the precedence given to flying the Royal Mail could invert the social order. At Kisumu in 1938, the *Imperial* aircraft in which the Duke of Gloucester was scheduled to return to London was reassigned to fly mail to Durban when the aircraft intended for that flight was taken out of service – the mail certainly got priority, but so too did the few ordinary passengers booked with the air mail.[16]

In January 1938 *Imperial* issued a nine-page booklet entitled 'The Public and Ourselves'. Signed by Woods Humphery, it urged 'correct behaviour' by crews, especially as regards dress, speech, courtesy and punctuality. The airline, it was said, occupied 'a unique position in the world': its services traversed parts of the world 'in which no other British organisation is represented in anything like the same way or with anything like the same responsibility'. Not only *Imperial*, 'but the efficiency and the manners and the character of a great many aspects of British life and of British aviation will be judged by our behaviour and by our bearing'.[17]

Having British men flying British aircraft on routes in the British Empire was consistent and patriotic. It was also a good return on investment in training, and meshed with the notion of a civil air reserve. It was a common view that flying British was also quality assurance. An aviation news magazine noted that the *Imperial* pilot Sam Wheeler (thirty-five) was 'a typical Briton'. His qualities were inner strength and 'depths of courage and tenacity'. The parliamentarian and entrepreneur, H. K. Hales, convinced himself that it was the nationality of the bloodied and shocked captain of the aircraft which collided with a vulture in mid-air over India that had saved his life. It was enough that 'a British hand was on the wheel'.[18]

Another *Imperial* passenger related his feeling of complete confidence not just in the organisation of the airline, but in the 'pluck and skill' of the pilots: two 'virile, alert young Britons' controlled and steered 'with such splendid spirit of quiet audacity and determination'. Introduced to an *Imperial* pilot at Heliopolis, Cairo's air station, W. J. Makin found that the vivacious man 'with little wrists of steel' was likeable and dependable. Having survived a crash at Gaza in 1930, the parliamentarian J. M. Kenworthy praised *Imperial*'s quiet, plucky Cockney wireless operator: the salt of the earth, he leapt from the wreckage to extinguish the landing flares which might have set fire to the aircraft. He saved both Kenworthy and the Ibn Saud's Foreign Minister, who was making his first flight.[19] Flying from Johannesburg to London in August 1933, a Durban resident noted the British-built aircraft and 'had a good look at the pilots who were also of British make and seemed capable of carrying out their important job during the journey'.[20]

Pilots' uniforms conveyed professionalism. But dress was not just a matter of pleasing passengers. Even in the absence of any passengers on the 1933 *Imperial* survey flight over the African coastal route south from Kisumu to Durban, the captain was perturbed to find his crew without a standard uniform. He instructed a local tailor to sew them white overalls and khaki tunics and shorts. Englishmen had to be correctly attired, whether in public assessing the availability of fuel, launches and hotels, or alone while surveying the weather, the water depth, currents, tides and mooring facilities in sheltered bays.[21] Dress – and styles of address – reinforced the social hierarchy in and outside aircraft.[22]

The Empire's diverse dress codes stuck in the memory of one *Imperial Airways* pilot. Fifty years after he stopped flying, John Spafford recalled that wearing a dark lounge suit for dinner at Kisumu

Figure 10 Authority and service: *Imperial Airways* stewards flank, left to right, radio officer, captain and first officer, 23 July 1937.

was overdressing, whereas dinner suit and bow tie were *de rigueur* in Karachi. Conformity was a social touchstone in other ways too, as Spafford sensed. While flying on *Imperial*'s India service he was made an honorary member of Karachi's sports clubs, but he felt unable to use them to return the private hospitality he received from the city's wealthy Indian entrepreneurs whom he met on his aircraft. In spite of progress in aviation and in many other fields, he wrote, the Empire stuck to Victorian traditions: if he had been seen 'hobnobbing with the natives' in a 'White' space he would have been blackballed. An innocent house-visit quickly became discreet socialising. Spafford was puzzled. Poorly educated and self-consciously working-class, he realised it was only his work for a prestigious airline which let him pass through the 'sacred portals' of the Karachi clubs.[23] He assumed wrongly that occupation could submerge colour as well as class. Spafford registered embarrassment but not shame.

The sense of imperial (un)belonging played out in other spheres. The constitution of the Empire airline forbade hiring pilots from beyond Britain, but even if it had not, employing men from the colonies and dominions might have struck many as 'just not right'. Lord Montagu's opinion of 1919 that the Government of India should not be allowed to refuse licences to Indian pilots or to prohibit Indian ownership of aircraft was enlightened, but it did not get to grips with the distinction between pilots and planes on internal and overseas air services. A more conservative sentiment was not aired much. When it was, it could land the speaker in trouble. In the late 1920s, Lord Chetwynd, the *Imperial Airways* Director who was being groomed to succeed Geddes, was persuaded to resign on grounds of ill health after his insulting innuendo about the ability of Indian pilots. In the same year, the Indian Government did in fact send three Indians to England on aviation training scholarships. Their subsequent appointment as aerodrome officers rather than pilots confirmed the suspicion of an Indian council of state delegate that the British would accuse Indians of lacking air-sense just as they were reputed to lack sea-sense. Tentative de-racialisation of British air enterprise was apparent in 1939 when *Imperial* first selected two Indians for training as first officers: they were to crew its jointly owned Trans-Continental Airways on its confined regional services.[24]

Like the Australians who in 1932/33 threatened a boycott of Indian pilots,[25] the outspoken C. G. Grey had firm opinions on people other than Britons flying British passengers and cargoes. Years after the Chetwynd episode, he used a letter received by the *Imperial* station manager at Salisbury to make his point. A 'Nyasaland Boy' had written requesting employment as an assistant pilot. His eagerness to 'drive' a machine – which he spelled euphoniously both 'Aroplane' and 'Heroplane' – affected Grey like an emetic. His editorial vomited about 'niggers' limited cerebral capacity' and their unsuitability to flying. Grey abided by his motto that writers should always sacrifice a fact for a memorable phrase. The Nyasa's letter evidently circulated with some amusement. When it fell into the hands of the London *Times*, the paper reprinted it, ostensibly to show the fascination exercised by aeroplanes on African minds and their 'perfect faith in the simplicity of the white man's inventions'.[26] The reprint did more. It belittled. It betrayed a confidence. It poked fun by denigrating people who at least tried to use another language. As the Nyasa's letter followed a previous unanswered request for work, its publication also disclosed even more disregard.

Imperial's flying crew had felt disregard differently, airing this at least once via the airline's 'Captains' Committee'. The captains might have been speaking for all aircraft crew when, in 1938, they voiced anxiety over job security in view of the Air Ministry's new physical fitness requirements. They also expressed concern about exposure to tropical climates and diseases, and about long periods away from home. Their request was for a £200 salary increase in compensation. *Imperial*'s response was cool: fitness requirements were unchanged;

pilots flying the Empire routes spent more time in temperate climates than in the early 1930s when they were based in tropical or sub-tropical climates; captains spent less time away from home than ships' captains; flying was less stressful than in the days when captains flew in open cockpits without wireless or navigation aids. The captains' wish to earn a fixed salary in place of basic pay plus a service-linked element also had a cool reception.[27]

Cabin crew

Aboard *Imperial* aircraft, male cabin attendants (called stewards) attended to passengers' needs. By 1935, twenty-six of them worked the 21,000 miles of Empire routes. They dressed as ship's stewards, wearing white jacket, black bow tie, wing-collar shirt and peaked cap.

Stewards' roles were diverse, assisting passengers in multiple ways and managing cabin use for the uninitiated. Ordering, loading and serving food was one task, albeit that on the first generation flying boats the stoves on board could be used by the valets of valued passengers.[28] There was literally no room on board for apprentices; only experienced 'hands' were hired. Several years on a leading steamship line were expected, and also, if possible, a period of service in a Pullman dining car. Stewards had to be accustomed to working in a confined space. Catering, housekeeping and budgeting skills were necessary too: supplies had to be bought from local shops and the open-air market in Croydon. Meals were not any slap-dash, picnic-box affair. Steward Jefcoate noted six meat dishes on the luncheon menus, each one served with care and attention comparable to a West End hotel. Only the crockery was inferior, composition being used instead of china in order to save weight.[29] A 1930s cigarette card in the John Player series of fifty international airliners gives a glimpse of catering aloft. Using a gaudily touched-up photograph from a *Weekly Illustrated* magazine, the card shows a uniformed steward preparing a salad lunch in an aircraft galley boasting iceboxes and wine-bottle bins.

Refinement, deference and politeness were essential elements of cabin service. A steward who had previously worked for the Cunard shipping line recalled being ordered to address every passenger by name. Failing that, a title such as 'Sir' sufficed. Stewards were also expected to be smart and well mannered. One unattributed claim is that attendants who went aboard with a loose button, or who coughed or sneezed even once, were put on the proscribed list, demoted or dismissed.[30] The list of imperial etiquette stopped short of insisting on confidentiality, presumably because engine noise made it difficult to overhear conversations.

Being sky-butler to the elite no doubt called for patience, but not necessarily more so than when some of the Empire's untitled mortals stepped aboard. Jefcoate remembered that stewards had to deal with an 'almost embarrassing' variety of people, ranging from king to prime minister to clerk. Stewards, he insisted, were not concerned with their rank. Each passenger's needs had to be studied and satisfied. Pleasing such a variety of people was not easy, but there was one sure guide: everyone expected and got the best.[31] Some were even roused to report it. Writing to the *Daily Express* about a flight from Nairobi to London, a contented passenger wrapped admiration in a reference to helpless, fragile infants. Apparently, if one handed a nine-month-old baby to the air steward he would bottle and bathe it at the correct times, and it would be chubbier when it arrived at the journey's end.[32]

Imperial's male nannies were gentle men indeed. They had a lot of practice too, for the airline was slower than many of its American and European counterparts to employ airhostesses. Yet the men also had to fit the image that they were more useful than women in assisting crew at aerodromes and during flying emergencies. Dealing with airsick and obstreperous male passengers was apparently one such crisis. Even so, some of a steward's tasks were part of what was once thought of as 'women's work'. A 1936 in-house airline magazine identified these attributes: experienced and consistent; part psychologist and part actor; methodical and meticulous; small, agile, quick moving, deft and proud.[33] Passengers appreciated imperturbability and selflessness. After *Horsa* made a forced desert landing in 1936, the 'marvellous' steward insisted on 'ministering to passengers as though nothing had happened'.[34] This was indeed 'emotional labour'.

When the Empire-class flying boats entered service in 1937, *Imperial* instituted a new category of cabin crew to manage paperwork that included recording passenger and passport details, checking manifests of luggage, mail and freight, and handling customs documentation. Preserving the maritime tradition, the 'purser', as he was called, was also required to operate a fire extinguisher (by perilous wing-walking) if an engine ignited when starting. He also had to prepare flight reports (using an on-board 'Baby Empire' typewriter). One ex-purser recollected that the task was a formality easily abused: superiors rarely read the reports before filing them.[35]

Many men were attracted by the prospects of flying imperially as a purser (the position was renamed 'flight-clerk' apparently to minimise pay claims). Press advertisements for forty positions in 1936 attracted 2,600 applicants.[36] One successful candidate was a twenty-seven-year-old Sydney man formerly of the P&O and Cunard White Star line.[37] Another appointee recalled that selection could be less meticulous than publicity claimed. The commercial assistant to *Imperial*'s traffic manager (previously the duty officer at London's Croydon airport) recruited Ian Driscoll after only a perfunctory 'interview'. Serving on the India sector, he earned a base weekly wage (£3 1s 4d) and earned extra (10s) for each day's flying, unless it was east of Calcutta in which case he was paid an additional shilling. Only men who loved aeroplanes and adventure could endure the airline's parsimony, Driscoll recalled. Doing stewarding as part hobby and part work left the airline in a strong bargaining position. Driscoll's own endurance was stretched after an incident when *Imperial* punished pursers collectively for a smashed chandelier in a Singapore hotel. In an 'inequitable affront' they were relocated to an inferior 'Eurasian' hotel.[38]

In Cairo, *Imperial*'s crews stayed in a villa. Among them was Hilary Watson, who joined the airline in 1934 aged twenty-four and spent his first weeks washing out soup thermos flasks in a kitchen at Croydon. Before being promoted to steward on the Mediterranean sector, he was station manager at Brindisi. There, some passengers (badly behaved and so rich they couldn't do anything for themselves) asked for directions to a brothel as soon as they landed.[39]

A sign that *Imperial Airways* was relenting on employing only men for work in or around aircraft was its appointment in February 1939 of a supervisor of women employees. The appointee, a woman who had been country branch secretary of the Women's Volunteer Reserves for the first eighteen months of the First World War, filled the post briefly before she switched to the Ministry of Munitions.[40] Hiring practices at other airlines were less sexist but not necessarily more efficient: over a twelve-month period in 1937/38, half the twelve air hostesses hired by the Dutch airline, KLM, had left to get married.[41]

Ground crew

Airmen acquired at least some of their aura in relation to the less glamorous work of lower-grade airline employees. Londoners, and those arriving in the imperial capital from elsewhere, would often have their first encounter with *Imperial's* non-flying staff at Victoria railway terminus. There, the young pretender to imperial transport squeezed into a corner under the ample roof of the established nineteenth-century overland transport giant.

If not inside the terminal building, then soon after, departing passengers would experience the trappings and pomp of Empire. In the early days, a uniformed pageboy was in attendance on the coach to Croydon. At the airport, the station superintendent greeted travellers. His sartorial trimmings included peaked cap and white gloves.[42] The parade was not to everyone's liking. Grace Crile found the experience cloying and patronising. Recoiling, she said Victoria was like a kindergarten. At Paris the airline's cringing steward struck her as the personification of rank- and rule-bound England. Unwilling to use his discretion over who could board an airways bus, he would listen only to the aircraft captain. 'Thus is England governed', she scoffed.[43] Thus was the Empire flown. On outbound flights from Croydon, a familiar social order was compensation for being hoisted away from mother England. Inbound, it was the past retrieved. The same was true of language and diet.

British ground crews and support staff scattered around the Empire's airfields were probably reminded often about being representatives of their home country and its culture. Flagged aeroplanes and their flying crew came and went often, but ground personnel were more rooted. Screening even for un-elevated Empire service was thorough. The 'energetic young men of the right type' whom *Imperial Airways* selected as its first overseas station superintendents were handpicked from among short-service officers in the RAF and from public schools. Once a nucleus of staff had been trained in London, a more formalised and rigorous system of annual recruiting and training was followed. Applicants had to be between seventeen and twenty-one years old. Groundlings were expected to be matriculated, healthy and able to adapt to life abroad. It was an unspoken assumption that only unmarried British males would be recruited. Achieving global pre-eminence would be a singular preoccupation: at the start of overseas services, *Imperial* forbade its station managers to marry for three years, even though this prohibition threatened the propriety of stable, monogamous relations among colonials. It was not until 1939 that the airline hired fourteen British women already living in the locality of Empire airfields to act as 'matrons'. Their job was to look after children at overnight stops on the African and Australian routes.[44]

Each year, a panel sifted through a shortlist of forty male candidates suitable for service at overseas airfield ('stations') where they played a crucial role in the web of imperial flying. Selection criteria included not just ability, but also character, appearance and residential address. The classier contenders – not just the lads from Croydon – were sent for assessment by the National Institute of Industrial Psychology. This was no old-school-tie affair; this was scientific selection for Empire service. A 'psychographic profile' rated each candidate on a three-point scale of intelligence, scholastic attainment, literary ability, constructional ability, sociability, sympathy, self-confidence, aggressiveness, leadership, carefulness, persistence, emotional stability and physique.[45] It is unknown which attributes were used as predictors of celibacy or unsuitability to marriage.

The six highest-scoring applicants began a three-year stint of paid in-service training each year. Between 1929 and 1939 approximately seventy-five were recruited. The first two were the seventeen-year-old John Brancker (son of Sir Sefton) and the eighteen-year-old Keith Granville. Both rose through the ranks. Brancker became an overseas regional manager before he took up the posts of Traffic Director and then General Manager (International Affairs) with *Imperial*'s successor, British Overseas Airways Corporation (BOAC). From 1953 to 1960 he was Traffic Director of the International Air Transport Association (IATA).

Granville became BOAC Chairman in 1971, forty-three years after starting as an apprentice clerk in Airways House, Charles II Street, London. Following brief postings in Italy, at the age of twenty-one he was posted to Dodoma, Tanganyika. There he operated on a loan from the local hotelkeeper for ten weeks before the first (long-delayed) African service brought his wages. Stints at Bulawayo and Broken Hill were followed by home leave (by sea). Granville's next postings were at Alexandria's Ras-el-Tin flying boat base and the new land plane airport at Dekheila. It was a career ignited in his final school year at Tonbridge in 1928 by a lecture from the Air Ministry's Wing-Cdr Blake on the future of British imperial air transport. After his mauling by aviation's senior men when his round-the-world flying ambitions ended in India in 1922, Blake is unlikely to have entertained his audience with bravado. Rehabilitated, he gave a talk that persuaded Granville not to sit entrance exams for the Royal Military Academy at Sandhurst. Self-deprecatingly, Granville recalled that his backup job application to ICI failed because, although he batted well, the firm's cricket team needed a slow left-hand bowler.[46]

Another early recruiting success was Ross Stainton, who joined *Imperial* as a trainee in 1933. In 1936, aged twenty-two, he started work as an overseas station officer. Over three years he was assigned to Italy, Egypt and the Sudan. In a 1996 interview he recalled passenger types, work routines, station staff (engineer, accountant and office boy to 'cool the beer'), climatic conditions, leisure activities and chasing animals on landing grounds. He remembered unpalatable breakfasts served aloft, and casually stashed gold consignments whose weight bent the aircraft skin. The youthful recruit, it transpires, had no boyhood fascination with flight. For Stainton it was still the sea that beckoned. Rejected on medical grounds by the Navy, he approached *Imperial* at his mother's suggestion. He showed no sense of being part of either a temporary or a momentous, irreversible shift in Empire service. After the Second World War his career in British civilian aviation progressed. He was appointed Managing Director of BOAC in 1971 and was knighted after retiring as Chairman of British Airways in 1980.[47]

Brancker, Granville and Stainton vindicated *Imperial*'s hope that the select few who entered civil aviation when its traditions and standards were being established would be 'of a type adequate to build the prestige of British air transport in all parts of the world'. The *Imperial Airways* agent at Malakal, Mr Grey (the 'nice chap from the university' whom the District Commissioner mentioned in a letter home in 1937), would have been one of these types. So too was Richard Poland, a graduate of the Royal Naval College at Dartmouth, who joined in 1932 as an eighteen-year-old traffic trainee. Unfortunately for *Imperial*, he took his first and only airline job with another British carrier in 1934 before he too advanced to eminent aviation and public service. Aged twenty, Gilbert Lee was recruited by *Imperial* in 1931 from a college in a Nottinghamshire coalfield town. He also served abroad before taking on senior airline management positions after the Second World War.[48]

Fresh out of school, Geoffrey Pett, a Kent policeman's son, joined *Imperial*'s Commercial Trainee scheme at the same time as Stainton. While waiting for his first interview at the airline's offices at Victoria Station, he felt that he stood no chance competing with the likes of the Eton boy wearing a morning suit and top hat. After sitting 'stupid exams' at the Institute for Industrial Psychology, he was finally accepted into *Imperial* in September 1933. Besides himself and Stainton, *Imperial* took on three other English boys (one was Douglas Grey, nephew of Lord Grey) and two Indian boys.

Trainee pay for the first year was 15s per week, doubling in the second year and rising to 45s in the third year. Pett's three-year 'apprenticeship' started at Croydon and Victoria in the offices of the General Sales Manager, the Traffic Manager, and in the accounts, meteorology, and foreign divisions. His first experience of overseas service was at Brindisi, where he met passengers off the 3.30 a.m. train and transferred them to the hotel and their baggage to the flying boat. When, aged twenty-one, Pett started work as a permanent airline employee in September 1936, his first job was to load the new Empire flying boats for stress and trim tests at Short's yards in Rochester. Just before Christmas 1936, he flew with the festive season's mails on *Centaurus* to Alexandria.[49]

'Men of a new calling' was how one commentator described *Imperial*'s overseas station masters in 1934. Fifty of them at the time, their work was likened to pioneering: they worked in lonely deserts and in the hearts of forests, cut off by rains when 'native roads' became impassable. The conditions called for diplomacy, initiative and judgement. Passengers had to be studied and made comfortable; local chiefs had to be dealt with shrewdly and courteously.[50] C. G. Grey wrote that the Superintendent at Lydda (Palestine), Mr J. Ramsay, was said to have 'a natural facility for dealing with Worthy Oriental Gentlemen' and the knack of getting along with 'the best sorts of British of all grades'.[51] In Africa, delicacy was called for after flying boat mooring buoys started disappearing at Malakal, Kampala and Kisumu. Crocodiles, it seems, were partly responsible. They mistook the red rubber for raw meat. The sunken buoy retrieved from Lake Victoria with half a dozen spears sticking in it raised other questions. Africans may have targeted it to protect crocodiles regarded as sacred, or its presence may have been associated with some local misfortune. 'Patient explanations by the airway staff, however, mollified the natives.'[52]

Romance, exaggeration and danger aside, Empire airfield superintendents were mainly aerodrome managers with responsibility for handling ground services, freight and mails. Keeping accounts and taking readings from weather balloons were subsidiary tasks. Managing people may have been more difficult, not least at remote outposts. Even at small, remote Lumbo in Mozambique, the staff complement in 1939 was two traffic officers, two engineers, two Portuguese coxswains, ten African seamen, a cook and a 'houseboy'.[53]

At remote Empire airfields and flying boat bases, the superintendent was also expected to be a hotelkeeper. He needed to be familiar with catering and business. He also had to be tactful and patient dealing with travellers (the 'most unreasonable of mortals') in low-latitude climates that shortened the tempers of host and guest. Linguistic skills helped considerably; this meant acquaintance with mainstream European languages rather than African or Asian speech. Skills in spoken English would have been expected of the indigenous Indian, Egyptian and Sudanese wireless operators hired by colonial public works departments to service the new airways.

Pett's recollections (recorded in 1995) offer a glimpse of tasks that went beyond daily servicing of passing aircraft. He himself recalled checking river moorings at Malakal for crocodile damage. It was a task he undertook while at Juba toward the end of six years' service in East Africa before war broke out. His real *Imperial* service started in Nairobi. It was the first of several postings: before turning twenty-five and returning home to England on leave in mid 1939, he had been Station Superintendent at Mbeya, Lindi, Butiaba and Juba. Among his first tasks at Mbeya was allocating rooms to passengers and crew. He was assisted by the station clerk, an Indian – 'all the Indians were called Singh'. The task was tricky on the rare occasion that there were several women passengers and when two full planes stopped overnight. Getting a naked, hung-over woman passenger dressed and aboard an early morning flight required some stealth and strength. Pett was on his own taking the mail by car into Mbeya, the first time he'd ever driven.

Figure 11 Different not uniform: *Imperial Airways* ground staff handling a flare buoy at flying boat base, Kisumu (Kenya), 1938.

Another first was shooting buck and serving venison at the rest house as a way of trimming catering costs.

Back at Nairobi, Pett worked on plans for provisioning flying boat stops across Africa. His work stood him in good stead for creating a 'station' from scratch at Lindi on the Tanganyika coast, including mapping deep water, positioning moorings, and supervising construction of a jetty and a small terminal building. He flew there in a hired Wilson Airways light aircraft and waited the arrival of stores by boat. The coxswain, Bob Percival, an East African fluent in Swahili, also arrived by boat. Pett remembered good relations with Africans, and amazement at their limited material resources. The single quaint technological incident he recalled involved an African who complained that the radio aerial in a tree was killing it: the crackling (electricity) sounds were its dying screams.

Pett left Lindi sadly, but none too early, having caught malaria six times. His next posting was to the emergency landing ground at Butiaba, where his job was to close the facility and move all the equipment (and the camp five miles away) to a new base near the established landing ground at Juba. The massive operation involved a three-day river journey to Nimule, towing the large Shell refuelling barge and carrying the passenger launch, patrol launch, and radio and office equipment. From Nimule it was a slow 100 mile overland journey to Juba. The stretch of water used by the flying boats was some way down river at Rejaf. Pett supervised clearing of a path and road (by prison labour). He oversaw building of a corrugated iron and wood airport building comprising engineering store, customs office, baggage bench and baggage examination area, waiting area, airline office and radio room.[54]

Imperial's management must have expected that wherever their employees were stationed they would be sustained by an ethos of duty, service and teamwork redolent of established British imperialism. At *Imperial*'s eighty 'ports of call' in 1938 its employees were the face of the Company. More than that, as General Manager Burchall noted in the same year, the airline's 3,000 employees in thirty countries were fortified by the knowledge that their work was 'of national and Imperial importance'. In the circumstances 'stationmasters' might have found it hard to resist an exaggerated sense of self-importance. Like previous imperial formations, twentieth-century aviation presumably was not inimical to the pompous 'Pooh-Bah' who held several offices at once and fulfilled none of them.[55]

Distance from the restraining influences of home, neighbourhood, Church and beat-policeman was a perennial problem in the Empire. Missionaries, soldiers, administrators and settlers had succumbed. Now it was the turn of the birdmen. A Church official remarked once that Englishmen deteriorated rapidly when far from home; George Orwell knew the phenomenon from his five-year spell as an imperial policeman in Burma in the mid 1920s, and worked his experience into his 1934 novel *Burmese Days*. He could not have imagined that air transport might help arrest personal decline and disillusion with

Empire by making shorter-term periods of overseas service feasible. The Australian pilot Lores Bonney recommended maximum year-long postings for 'the particularly fine type of Englishman' whom she encountered in the African colonial service. A known and fixed term of service helped to combat alcoholism, 'the kind of deterioration the British Government would not tolerate'.[56] The morale and morality of the imperial order were at stake.

The dominant heroic image brought by pilots to Empire aviation could be seriously undermined by the debauchery of just a few British civil aviation personnel overseas. In one instance events were grave enough to be spoken about in the House of Commons. They were grave enough to refer for investigation, and possibly even serious enough to dismiss as rumour. In 1938 an ex-Under Secretary of State for Air alleged that an *Imperial Airways* operating manager (at Karachi) had been excluded from seventeen clubs because he had been drunk and had insulted (and once, sexually harassed) the wives of pilots. Investigating these allegations, Sir James Price, a Director of *Imperial*, found only gossip, specious and contradictory evidence, and misunderstanding (the official had withdrawn his application to one club on a technicality).[57] In April 1939 when the Iraq area manager told Burchall about a passenger's complaint of drunkenness among pilots at night stops, Burchall dismissed the allegations as exaggerated, disloyal tittle-tattle. Later, in December, he claimed coincidence and mal-inference when the same area manager reported that the captain involved in the *Calpurnia* flying boat crash at Lake Habbaniyah in 1938 had been spotted at a late-night cabaret in Alexandria the previous night.

Defending pilots against accusations of excessive drinking started at least as early as 1932. At the time, the Deputy Director of Civil Aviation reported that the case of drunkenness at Cairo involved a wireless/telephone operator and not a pilot as was once supposed. Making no mention of safety, he stated indignantly that rumours of drunken pilots damaged the goodwill of *Imperial Airways* and were unfair to an excellent body of men. Yet, during his six-week Empire airway inspection tour in the summer of 1939, *Imperial*'s Operations Manager, Mr A. C. Campbell-Orde, personally saw an off-duty and out-of-uniform *Imperial* captain the worse for drink in a Kisumu bar. Whereas Campbell-Orde used the episode to say that it would be a 'godsend' if the airline provided a crew mess in out-of-the-way places where there was nothing to do but loaf and drink, a sharp letter to Burchall from a senior airline colleague in December 1939 pointed out that another in a series of observations only added to the reputation which *Imperial* captains had of being heavy drinkers.[58]

Degeneration came to the attention of Maj. H. G. Brackley, *Imperial*'s

Air Superintendent, soon after he arrived in 1929 at Heliopolis. Brackley was first engaged by the airline to sort out the birth pang pilot strike in 1924 (aviation's selective adoption of maritime terms never included use of the word 'mutiny'). Subsequently his work involved route planning and development. On this occasion he substituted temporarily for the regular station manager. Brackley, whom a pilot remembered as 'always calm, always immaculate, and extremely proper', wrote in a letter home that one of the ground staff had become quite 'woggish'. He dressed as a native, hennaing his feet, and was worshipped by a native girl with whose family he spent most of his time. It is unclear whether Brackley took the offence personally, or whether he construed it as an offence to the Crown, the Empire or the airline. To offend one may have been to offend all. Whatever the case, the cure for such fearsome decline into immorality was three months' leave in England. There, old habits and ideals could be restored. The subordinate – 'such a nice fellow' – had not grasped that certain things were just not done in this life. Brackley would have been appalled by the peccadilloes of the French Air Sergeant whose barefooted, bare-breasted and wild-haired Arab mistress was on view in the mess hall at Timbuktu airfield.[59] Indiscreet display is precisely how John Bull commentators would have distinguished the French and British air empires.

Concern about expatriate behaviour and reputations when off duty could be compounded by anxiety about the agents who represented and acted for *Imperial Airways* overseas. At Moshi, a minor Tanganyikan airfield, the local hotel proprietor acted for the airline in 1934. *Imperial*'s East African regional manager, J. W. S. Brancker, was unimpressed. The employee's energetic solicitation of traffic had to be weighed against his keeping poor tables and rooms, being 'an incredible liar', being slow, careless or dishonest in remitting fare receipts, and charging incorrectly for excess baggage. His dress did not redeem him. Not having an airline uniform, he wore a grey flannel suit and large white toupee with brass badge when he met aircraft. Brancker thought the outfit made him look like 'a municipal dustman'.[60]

Appearances counted in Iraq too, and at senior level. After his visit in 1939, *Imperial*'s foreign public relations officer felt that the Area Manager, Maynard Marais, was 'unable to control British staff and retain their respect in a country with a native population and a strong colour prejudice'. The airline's Operations Manager described him as intelligent and active, but was equivocal about the significance of Marais being 'neither English by appearance nor breeding'. Nevertheless, being anxious also about the limited respect and contacts of *Imperial*'s Near East manager, Gen. Huggins, he advised that all area managers should be more carefully selected, preferably retired RAF, university educated and of high social standing.[61] The ideal type, presumably, was a man like Air Vice-Marshal Sir T. Webb-Bowen who, after RAF service, was Staff Manager at *Imperial* for about four years before he resigned aged fifty-eight toward the end of 1937.[62]

In time, *Imperial* relented on its initial preference for hiring unmarried men to staff its overseas stations: solitude may have undermined perceived social pedigree. But the 'wrong kind' of marriage could be fatal. The airline's Cornish agent and wireless operator was apparently banished to remote flying boat bases first at Gwadar and then at Jiwani in Baluchistan because his marriage to an Indian woman contravened the rules of British community life in the East. The couple and their many children started life at the new Jiwani base under canvas. Similar disapproval may have explained the stationing of a retired Siamese naval officer at an *Imperial* emergency stop away from Penang.[63]

Another category of concern about airline ground operations involved work practices at remote Empire airfields. Loose cannons could be disruptive and potentially expensive, as a 1929 incident proved. Even under provocation, Mr Johnson, the Assistant Manager of the Mesopotamia Persia Corporation, was unwise to have hit a Persian wireless operator at Bushire who once refused to work when an *Imperial* airliner was in the air. Johnson was transferred to Basra pending police submissions to a court. The British Government's representative in Bushire expected the case to be dropped.[64] 'Loose characters' were also troublesome. The British Political Agent in Bahrain, Col G. Lock, and the Residency Agent at Sharjah, were drawn into airfield affairs in 1937 when the Station Superintendent, Alastair Thomson, twice sought missing cash. A petrol tally clerk (an indebted drinker and opium smoker who lived with a prostitute and had a prison record) and 'refuelling coolies' were variously lashed, imprisoned and deported to Iran.[65]

A year after leaving his post as Kenyan Governor in 1930, Lt-Col Sir Edward Grigg raised anxiety about expatriates employed by *Imperial Airways* at its African bases. He remained sceptical about aviation for twenty-five years: in his day, he recalled later, visitors who went to Kenya by sea stayed longer and saw more. Flying, he said, had done nothing to enhance understanding between nations and individuals. Speaking in 1931 during discussion of Woods Humphery's address to a Royal Society of Arts meeting, Grigg warned that maintenance crews would have to be watched and looked after carefully. Something happened to mechanics in Africa, he sighed. Unable or unwilling to be more specific, he added only that the men did seem to be less reliable there than elsewhere.[66]

Grigg's audience would have been left wondering whether Europeans succumbed to climatic lassitude or whether African assistants infected their work ethic. In the lecture hall, thoughts may not have strayed far from those of the American private pilot who, ten years later, drew on his Latin American and African experience in recommending six-monthly tours of duty by mechanics from temperate zones. A medical doctor, his view was that growth and expansion in the mechanical age had demonstrated clearly that complex machinery was meant for only certain races. Few peoples, he said, possessed 'the special ability, the patience and thoroughness, the inborn sense of ultimate responsibility, and that peculiar comprehension of things that go round'. And those that did, were not people of the tropics.[67]

Doubts about the aptitude of indigenous labour for aviation-related work arose in British minds in the 1930s in relation to servicing flying boat operations at overseas stations. Could local people be entrusted with using and supervising the use of marine equipment, including motor launches, moorings, buoys and flares? One view was that it was too risky to allow 'excitable' and 'negligent' natives to delicately manoeuvre high-power motorboats when patrolling alighting areas in the teeth of difficult winds, currents and tides. Not only seamanship but also marine signalling skills were in question. Several pilots' unhappy experiences with indigenous watermen on flying boat expeditions were not cited as evidence, but failed experiments later with workers in Portuguese East Africa were. A colonial official in India who complained about a dissolute coxswain remarked simply that 'Eurasians lose their heads'. Rangoon's Irrawaddy Flotilla Company declared that 'the mind of the lower class Burmese is not remarkable for quick thinking or rapid action in an emergency'. Anxiety also arose about illiterate people fathoming rules written in the English-language marine manual issued by the civil aviation directorate in London.

From 1936 the Air Ministry, the Treasury and *Imperial Airways* considered engaging only expatriates as coxswains. The affordability of sending British Navy ratings to work at every one of the thirty-two flying boat stations in the Empire was a contentious point. Home-trained coxswains earned more (£8 weekly) than locally engaged men (£3–£7), but passages, health insurance and accommodation had to be paid. The additional cost had to be balanced against the possibility of delays to services and damage to flying boats and launches. The additional annual cost of expatriates was estimated at four per cent of the capital cost of the launches. Practical compromise trumped racism. 'Home-trained' coxswains worked the bigger and busier stations, assisted by local Europeans and indigenous men. A division of labour may have applied too, with indigenous crews working the least powerful launches for ferrying passengers, flying crews, fuel and provisions.[68]

In India, the compromise arrangement evaded objection from the

Government (which wanted local labour used wherever possible) on grounds that it was the most consistent with operating efficiency and was temporary. But even appointment of coxswains in Britain did not necessarily involve selection of the most able men. A former *Imperial* member of staff at Southampton recalled that once negotiations between the airline and the Transport and General Workers' Union established that only registered stevedores and dockers could attend the flying boats in the harbour, he got to the front of the queue simply because he was a Union member.[69]

One explanation for mechanical ineptitude or carelessness among indigenous labourers went beyond climatic determinism and racial characteristics. It rested, instead, on the problematic status of both dark- and pale-skinned manual workers in the Empire. Once again, this was a matter that reached Parliament in London. Airline affairs shaded into Empire affairs. The Commons was told that in Egypt, rank and file *Imperial* employees (lower-middle-class 'John Citizens', like RAF mechanics) were known contemptuously as 'white wogs'. The derogatory abbreviation was used to describe low-paid White men who 'went native'.

The way to avoid consigning Britons to the category of poor White trash was simple. *Aeroplane* readers, most probably including *Imperial Airways* managers, would have seen C. G. Grey's advice to pay minor White officials and mechanics a special allowance so that they could live 'on an Englishman's scale'. The alternative was that they would disgrace their airline and country. Grey's proposal was as limited as his concern. His response to a parliamentary question about the low wages *Imperial* paid to its overseas Indian staff reeked of favouritism. The matter left him cold, he wrote. He was more troubled by the low wages paid to 'decently-bred Englishmen who have to keep up the dignity of the white man among a coloured population'. Indians, he claimed, were accustomed to working for low wages. They were not compelled to join or stay at the airline.[70]

As in other imperial industries, the ethnic labour record of Britain's Empire airline can be measured in more than numbers and ratios. Job descriptions, pay scales and privileges differed. Tasks were differentiated by skill – and therefore race. Employees wore different uniforms. The 1933 film 'Wings over Africa' contains a clip of a Sudanese man wearing a shirt bearing the words '*Imperial Airways*'. Caricature was perhaps an even more insidious discriminator. Shell's 'service station' manager at Entebbe aerodrome was identified, in colonial custom, without a title and by one name only. He was presented without the reassuring biography that organisations used to boast about all their other responsible staff. The man was said only to be 'about the best known and most reliable boy on the Cape–Cairo route'. A photograph of 'Yowsha' (captioned 'an imperial character') shows him sitting on the ground, balancing a clipboard on his knees and diligently making pencil entries on a Shell chart. His barefoot crew of six posed in their Shell-issue shirts and caps.[71]

The sight of identifiable, organised and loyal strangers would have reassured visiting passengers and aircrew. But uniforms, it transpires, were only partly about corporate image. The Africans employed by Shell to refuel aircraft were provided with uniforms both to give them 'a smart and workmanlike appearance' and also to 'enable the supervisor to recognise his five or six men quickly amongst the mob' that often surrounded incoming aircraft.[72]

Occupying perhaps the lowest rung of labour on the Empire air map were people who manually prepared and repaired landing grounds. Photographs show the activity into the mid 1930s. In one, a ragged band of twelve women clear Mayumba landing ground.[73] Another depicts supervised work at Li Rangu (Sudan). In a diary published almost fifty years after her African study tour, Margery Perham relates seeing construction of a small airfield in rural Tanganyika in 1930. Forty naked Wambunga were on site by six in the morning. In comments more typical of the White people she criticised, Perham noted how the labourers 'ceased to tickle the surface' and started putting their backs into their work when she appeared with her brother-in-law, the Provincial Commissioner. With his help, the men were soon 'shouting and singing, straining to carry trunks 100 yards over rough ground'.[74] Even the newest form of imperial transport depended in part on colonial labour.

Notes

1 NA, AVIA 2/422.
2 P. J. Lyth, 'The changing role of government in British civil air transport, 1919–1945', in R. Millward and J. Singleton (eds), *The Political Economy of Nationalism in Britain, 1920–1950* (Cambridge, 1995), pp. 65–97; F. D. Sprague, 'Imperial Airways Ltd', *Airway Age* (1928), 44–8; Warwick, 'The great trans-African airway'; *The Times* (15 November 1938), p. 22.
3 *Engineer* (12 June 1936), p. 623.
4 BAAM, AW/1/3949: Chief Engineer H. Hall to *Imperial* General Manager, 9 September 1936.
5 In his memoirs *From Fishing to Flying*, Allen Finch, an ex-*Imperial* navigator, recalled that whereas the carrier's pilots wore smooth barathea with gold rings of rank and a full-wing brevet, navigators 'had to be content with rough old serge, a single blue cloth band round the sleeve and no brevet'; www.rsne.com.au/af/chapter14.htm (accessed May 2011).
6 Brittain, *Wings*, p. 48.
7 *The Times* (29 November 1938), p. 16.
8 *Sphere* (4 March 1933), p. 292; *The Times* (16 February 1931).

9 W. S. Brancker, 'Commercial aviation', *Journal of the Royal Society of Arts*, 73 (1925), 649–63.
10 *Air and Airways* (April 1933), p. 12; Spafford, *Comings*.
11 *Air and Airways* (April 1933), p. 12; R. H. MacDonald, *The Language of Empire: Metaphors and Myths of Popular Imperialism*, 1880–1918 (Manchester, 1995), pp. 217–18.
12 Crile, *Skyways*, p. 232.
13 *Canberra Times* (2 June 1936).
14 B. Cassidy, *Flying Empires: Short 'C' Class Empire Flying Boats* (Bath, 1996), pp. 111–12.
15 Brittain, *By Air*, p. 60; Armstrong, *Pioneer*, p. 124; Morris, *Farewell*, p. 357.
16 SAL, MSB 750.
17 BAAM: Imperial Airways miscellany.
18 *Air and Airways* (May 1933), p. 54; Hales, *Chariots*, p. 94.
19 Webb, 'India to England in five days'; Makin, *Swinging*, p. 173; Kenworthy, *Sailors*, p. 249.
20 SAAFMY: Miller Papers, R43–2.
21 Jones, *Time*.
22 Cassidy, *Flying*, p. 110.
23 Spafford, *Comings*, p. 68.
24 Montagu, 'Aviation as affecting India'; Higham, *Britain's Imperial Air Routes*, p. 166; BL, OIOC, L/E/9/85.
25 See Pirie, *Air Empire*, pp. 155–6.
26 *Aeroplane* (8 January 1936), p. 48; T. James, 'Charles Grey and his pungent pen', *Journal of the Royal Aeronautical Society*, 73 (1969), 839–52; *The Times* (24 December 1935), p. 5.
27 BAAM, AW/1/3796: Notes on letter from Captain's Committee to *Imperial Airways* Managing Director, 28 March 1938. In 1939 a captain's basic annual pay was £475, augmentable by service to £1,300. The legal limit on flying hours was 125 hours in thirty days. Cassidy, *Flying*, p. 112.
28 Cassidy, *Flying*, pp. 123–5.
29 *Airways and Airports* (March 1934), pp. 333–4; Cluett, et al. (eds), *Croydon*, pp. 20–21; F. A. E. Jefcoate, 'Catering in the air', in L. Taylor (ed.), *Souvenir of Empire Aviation Day, 1935* (privately published, n. d.).
30 Hudson and Pettifer, *Diamonds*, p. 84.
31 Jefcoate, 'Catering', pp. 65, 94.
32 Reproduced in *Imperial Airways Gazette* (May 1933).
33 *Aeroplane* (19 August 1936); *Imperial Airways Bulletin* (8 September 1936).
34 *Sydney Morning Herald* (14 September 1936), p. 10.
35 I. H. Driscoll, *Flightpath South Pacific* (Christchurch, 1972).
36 Cassidy, *Flying*, pp. 119–22.
37 *Straits Times* (Singapore) (12 December 1937).
38 Driscoll, *Flightpath*, p. 87. Allen Finch recalled similarly perfunctory hiring practices; www.rsne.com.au/af/chapter12.htm (accessed May 2011). Later, as an outspoken member of the Association of Wireless and Cable Telegraphists, he refused to accept a posting to Egypt at £6 6s a week and was dismissed from *Imperial*; www.rsne.com.au/af/chapter14.htm (accessed May 2011).
39 T. Quinn, *Wings Over the World* (London, 2003).
40 *The Times* (16 February 1939).
41 *The Queenslander* (Brisbane) (18 May 1938), p. 31.
42 Cluett, et al. (eds), *Croydon*, p. 23.
43 Crile, *Skyways*, p. 18.
44 Brenard, 'The romance of the air mail'; R. Brenard, 'Life along our Empire airways', *Aero and Airways* 2 (December 1935), 103–5; *Aeroplane* (31 May 1939), p. 708.
45 A. S. Jackson, *Imperial Airways and the First British Airlines, 1919–1940 (Lavenham, 1990)*; Imperial Airways, *The Empire's Airway*.

46 *Imperial Airways Bulletin* (29 March 1974). In time Granville became IATA President. He was knighted in 1973 for his contribution to civil aviation.
47 BECM: Tape 377.
48 *Airways and Airports* (March 1934), p. 334; P. Gillibrand, 'Transport and communications', Paper presented to an Inter-faculty Symposium, Rhodes House, Oxford University, 1984; SAD, 767/7/22-23; *Dictionary of National Biography, 1986–1990* (Oxford); *Who's Who 2000*; *Who Was Who, 1897–1996* (Oxford).
49 Pett, *Beyond*.
50 R. Brenard, 'Men of a new calling', *Popular Flying* (November 1934), pp. 410–11.
51 *Aeroplane* (7 September 1938), p. 301.
52 *Time* (25 April 1938); *Evening Post* (Wellington) (21 April 1938), p. 8.
53 Sims, *Adventurous*, p. 114.
54 Pett, *Beyond*.
55 *Aeroplane* (19 March 1930), p. 494; Warwick, 'The great trans-African airway'; *Airways and Airports* (March 1934), p. 334; E. W. C. Sandes, *The Royal Engineers in Egypt and Sudan* (Chatham, 1937); *Listener* (20 October 1938), pp. 825–7; Imperial Airways, *Imperial*.
56 Gwynn-Jones, *Pioneer*, p. 147.
57 BAAM, N 30 (263): Minutes of Meeting of *Imperial* Directors, 17 January 1939 – Report from Price to Chairman.
58 Pudney, *Seven*, p. 126; BPMA, PO 33/3666; BAAM, N 532: Burchall Papers.
59 Bennett, *Pathfinder, p. 59; F. H. Brackley, Brackles: Memoirs of a Pioneer of Civil Aviation* (Chatham, 1952), pp. 298–9; W. Seabrook, *Air Adventure: Paris–Sahara–Timbuktu* (London, 1933).
60 BAAM, AW/1/7328.
61 BAAM, N 532: Burchall Papers. His name ('Marais') stamps the man as South African.
62 *Daily Mirror* (18 November 1937); Cassidy, *Flying*.
63 Driscoll, *Flightpath*, p. 82.
64 PDPG (October 1929).
65 BL, OIOC, R/15/2/1883 (B16).
66 E. W. M. Grigg, *Kenya's Opportunity: Memories, Hopes and Ideas* (London, 1955); *Journal of the Royal Society of Arts*, 79 (1931), 298.
67 Light, *Focus*, pp. 198–9.
68 NA, DR 9/68; AVIA 2/1013.
69 Southampton City Council, Oral History Unit, Item 19456. However, the informant's memory was flawed and in other respects he made errors: flying boat services did not start in 1938, more than two flying boats crashed and some passengers were killed.
70 *Aeroplane* (1 April 1931), p. 550; (22 March 1933), p. 488.
71 *Aeroplane* (2 September 1936), p. 307.
72 J. Glennie, *Personalities in South African Motoring and Aviation* (Durban, 1941), p. 135.
73 *Aeroplane* (2 September 1936), p. 307.
74 Perham, *East*, p. 227.

PART III

Virtual flying

8

Imperial plumage

Air services did not attract passengers effortlessly. Whatever the intrinsic appeal of air transport, it was easily undermined by the cost of leaving the ground and by anxiety. Airline advertising was about putting monetary costs into perspective, reassuring a nervous public, and building a passenger and cargo market from surface trips that could be superseded. From Britain's point of view, a prime task was to position its own Empire airline prominently in the public mind, and to persuade travellers and shippers about its qualities, notably its convenience, safety and comfort. Whether for travel or mailing, businessmen became a prominent target market. A 1936 *Imperial Airways* advertisement spelled out how appointments geographically far apart could be squeezed into a short time: the fictional ten-day African business trip did not include golf, but there was time for Thursday dinner with James in Alexandria, Saturday dinner with Gibbons in Juba and Monday dinner with Eric in Mbeya.[1]

Making the public 'airminded' was certainly part of deliberate acculturation in late imperial Britain. The hope was that flight would become part of a culture of modern mobility and a matter of everyday conversation and consciousness. Even those who never needed to fly, or who could never afford to fly, might have felt proud when reading about and seeing images of a great Empire airline: print, film and equipment displays in the capital and beyond could all stoke a glow. In view of the relatively few people who did fly on Empire services, advertising the new airway to and from Empire may have burnished imperial sympathies more than it boosted *Imperial* revenues.

Feathers

Imperial Airways crew, home airport and aircraft lifted the profile of the British Empire. In reciprocal fashion, the airline tapped the Empire for publicity. The publicity received and originated by *Imperial* also helped to project and remind people about the Empire between the two world wars. Advertising was about filling aircraft with passengers and freight, but creating an awareness of places in the Empire and embracing imperial inclinations were not coincidental. Indeed, they were intrinsic to the 'carefully thought-out propaganda' that C. G. Burge (editor of *Air Travel and Commercial Air Transport*) said would complement reliable air services in stimulating demand for air travel, and the transport of mails and freight.[2] In the air age, marketing *Imperial Airways* and marketing Empire were synonymous.

Starting slowly, but gathering pace through the 1920s, *Imperial*'s publicity department handled most aspects of airline promotion.[3] Some of it was cheap and facetious. Robert Byron thought that a series of 1929 advertising posters could appeal to neither wealthy passengers nor serious merchants. A 1931 poster design was surely no better. Intended to depict everything *Imperial* could show air travellers, it comprised a map cluttered to the point of illegibility with crude hand-drawn caricatures of the countries served or flown over. The outline of Australia, for example, contained a pair of cricket wickets, a rabbit, a Kangaroo and a sheep.[4] A 1931 'Summer Programme' advertisement for the Indian and Cape services listed the sequence of stops together with crisp, dramatic, clichéd encodings relating to legends and resources. Cairo: Nile and pyramids. Baghdad: jewel of the East. Gaza: gateway to the Holy Land. Basra: Sinbad, dates, oil and carpets. Juba: Nile headwaters. Nairobi: centre of big game country. Broken Hill: Katanga radium mines. Salisbury: King Solomon's mines at Zimbabwe. Bulawayo: junction for Victoria Falls and World's View. Johannesburg and Kimberley: gold and diamonds.[5]

These early advertisements did not sell speed. Rather, they sold places. More than that, they sold fictional places. *Imperial*'s aim was to boost ticket sales by temptation, using familiar images with which British people could readily identify foreign parts. An unintended outcome was fabrication of Empire. Accuracy was not the crucial consideration, as *Imperial Airways* ephemera show. The cover illustration on the summer timetable for the airline's European services in 1928 and 1929 was incongruous, premature and exaggerated. African air links were still a dream, yet even the European timetable presented an earthworm's view of an enormous lion standing imposingly in front of a fluffy white cumulus cloud suspended in an orange sky.

For its 1931/32 winter timetable to Egypt, Iraq, India and Kenya, *Imperial Airways* returned to pictorial design: an amateurish depiction of the two trunk air routes framed by the upper and lower wings, wing spars and fuselage of an airliner hurtling earthwards as if in a stunt dive.

It was one thing to have the Empire at one's feet, quite another to have it flashing before one's eyes as if part of a deathbed scene. The attitude was alarming, and the altitude from which one could gaze down at half the world was unattainable. Conspicuous features on the map were the Himalayas, the Arabian and Saharan deserts, and an elephant and lion each the size of the Atlantic sailing ship. In a tentative shift to using the work of colonial artists, the 1934 timetable folder had a cover design by the South African impressionist, J. H. Pierneef, who was then painting murals in South Africa House in Trafalgar Square.[6]

More pointedly than suggested by the lion image on the European timetable, other strands of *Imperial Airways* advertising cohered with an image of the British Empire as a male hunting ground, a vast resort for outdoor sport, relaxation and recreation. In 1932, in one of its many one-off advertisements, the airline sought to persuade readers of the *Field* magazine to fly to their shikar in India or Africa. In the previous year, a colour brochure used by *Imperial* to market its flying boat service to East Africa had featured a stereotypically drawn African with spear approaching a sprawling lion. The mane of the prostrate animal was outdone by the hunter's elaborate feather headdress. His armbands and earrings were presumably intended to establish something of his rank and authenticity. Three other African hunters emerge triumphantly out of forest in the background. Inside the brochure, a map of *Imperial*'s air route from Cairo to Nairobi was superimposed on a collage of warthog, buffalo, lion and rhinoceros. Headline text proclaimed 'Big Game in One Third the Usual Time' and 'The Direct Route to Big Game'. Textual patter emphasised that hunters who flew to the heart of Africa (via the 'blood-stained Sudanese battlefields with their memories of Gordon') could have a full week's shooting and return to London in less time than it took overland hunters to complete the outward trip. Previously, slow surface transport had limited Empire hunting to leisured men only. The new generation of hunters was encouraged to send trophies home by air.[7]

Imperial's publicity capitalised on sightings of game for passengers and on its ability to take people to jumping-off points for safaris. It was an enduring selling point. In 1934 the airline's manager in East Africa, J. W. S. Brancker, even recommended abandoning Moshi as a stopping point because Arusha was closer to the Serengeti Plains and a more attractive base for safaris. A 1936 edition of the *Imperial Airways Gazette* publicised arrangements with a reliable and experienced 'white' resident in Mpika district to meet shooting parties on arrival and to arrange safaris, including food, accommodation, African carriers and personal servants. In 1938 *Imperial*'s press department wrote to the Colonial Office requesting information about facilities for hunters and fishermen along its routes, and for advice about appropriate equipment and clothing.[8]

Spotlighting visual delights in the dominions and colonies by means of posters and decorated timetables was not the only way of advertising the Empire airline. Nor was it the only way in which public attention was surreptitiously drawn to the Empire in the air age. From 1929, airline self-promotion also included monthly publication of the *Imperial Airways Gazette*. It was available on request and, for a time, circulated as a supplement to the industry magazine, the *Aeroplane*.[9]

The *Gazette* was unavoidably imperialistic. Jingoistic slogans, such as the one carried in the July 1934 issue, boasted that *Imperial* offered 'the greatest air service in the world'. Apart from direct sales-pitches, the *Gazette* sought to popularise air travel and attract passengers by printing monochrome (often overhead) photographs of exotic places and publishing notes about leading personalities who had flown. Clever photography of aircraft probably did less to attract paying passengers than to portray the might and modernity of British airliners. Similarly, although the small number of people who took to the air would have actively scanned the gazetted timetables for the purpose of trip scheduling, many more people would have studied them abstractedly. The tabulated matrix of Empire places, and departure and arrival times, concealed the slender air corridors while accentuating the vast territorial coverage of Empire air transportation and, by implication, the extent of British influence and contact with the rest of the world.

Personal accounts of flights with *Imperial* made more interesting reading than timetables. Those published in the airline's *Gazette* included reminiscences of a journey from Rangoon to London, India to South Africa (the editor of the *Lucknow Pioneer*), and London to Baghdad and back (an *Aeroplane* journalist). Articles referring to the Empire airline and airway were also gleaned from foreign and regional papers. Extracts or reprints were taken from the *Nottingham Journal*, the *Sydney Morning Herald* and the Cape Town *Argus*, for example. An account of a flight to Entebbe by the M. P. for Penistone was reprinted from the *Yorkshire Times*. Articles written for other magazines by senior airline staff were also reprinted. One *Gazette* published the piece written by Managing Director Woods Humphery that first appeared in *Lloyds List*.[10] Talks were also reprinted, notably Assistant General Manager Burchall's lectures to two royal societies (the Central Asian and the Scottish Geographical) in 1932 and 1933. Copies of his lectures on air route politics were also available in *Imperial*'s publicity department at Victoria terminus.[11] Reprinting publicised the diversity of places where Empire aviation was under discussion and the variety of audiences who were drawn in.

Imperial sought print publicity beyond the pages of its own *Gazette*. In the daily and occasional press the airline's earliest advertisements have a sporadic, haphazard quality. Characteristically one off, they show little sign of a sustained campaign using either identical advertisements or variations on a theme. With few exceptions, advertising was not themed or timed to coincide with particular events. Advertisements were not placed strategically on particular pages of newspapers and magazines. The basis for placings in different outlets is unclear. If a wide reading public in Britain did form impressions about aviation and Empire from early *Imperial* advertising, they were incoherent.

The Empire airline got free news publicity in *The Times* but placed few advertisements in the paper. When it did, the focus was momentary. There was only one summer 1936 advertisement for *Imperial*'s African services: merging the head of an African with the outline shape of the continent created a highly generalised statement about destinations. The timing of such one-off advertisements could be curious. For example, although the airline's short advertising series in 1929 coincided with the seventh international aeronautical exhibition at London's Olympia, there was no mention of aircraft technology. Instead, *Imperial* chose to address 'Manufacturers, Merchants and Businessmen' with a message about the time and worry they could save by air mailing letters to Egypt and the Far East. The final line of text urged a visit to *Imperial*'s stall at Olympia, where two exhibits had an imperial reference. A slow-moving, painted panorama of scenes on the England–Egypt–India air route passed behind the windows of an airliner cabin section, and aircraft models were moved along on a map of the route in accordance with position reports from Croydon.

At the height of the fuss about slow Empire air services in 1934, a rare *Imperial Airways* advertisement appeared under the heading 'The Future is in the Air'. The ambiguity read as a bold-type admission that better service would or may still come. Curiously too, the bulk of the advertisement was a monochrome photograph of a young girl clutching a doll while being guided from one of the Company's airliners. When punchy publicity about the comparative speed of Empire air and sea transport would have been appropriate, the strongest visual allusion was to care and safety. As if by omission, the airline seemed to acknowledge that pace was not its selling point. Text at the foot of the advertisement did indicate that *Imperial Airways* travel was 'swifter, less fatiguing and more delightful than any other form of travel', but the text continued, and ended, by emphasising that air travel was so easy that 'small children often travel home unattended from abroad by *Imperial Airways*'. Later, the airline stretched its caring message by advertising the ease and speed of air travel as being 'particularly delightful for women, children and elderly people'.[12]

Eyeing a market among individuals intrigued by flight – and committed or converted to it – *Imperial Airways* placed most print publicity in the aeronautical press. Its series in *Air and Airways* in the early 1930s stressed luxury Pullman-design accommodation, excellent in-flight meals and the comfort (not to say convenience) of toilets in every airliner. The advertisements did raise the subject of travel speed, and the associated time saving for people travelling on business or sending freight. All passengers, of course, would have interesting aerial glimpses of historic and natural sites, and they were promised travel free of worry and trouble on 'entrancingly interesting journeys'. The advertisement headline, the name of the airline, began to carry the patriotic by-line 'The British Air Line'. The Empire and Crown loyalties of the airline never wavered; aircraft cabins were British firmament. Thus, on 12 May 1937 passengers in three flying boats heard the BBC broadcast of the Coronation service (probably indistinctly) while flying over the Mediterranean. One captain proposed a toast on the 'promenade deck' of his aircraft.[13]

One series of advertisements contrived to present the Empire airline in reassuring classical tones rather than in unsettling modernist tones. The series drew on a quotation from the 1926 English translation of *The Travels and Adventures of Pero Tafur, 1435–39*. It may have been unearthed originally by Sir Philip Sassoon, who used it as the epigraph for his 1929 book *The Third Route*. The tale of the twenty-five-year-old Castilian knight's wish to go to India from the Red Sea contained this most apposite sentence: 'And I finally concluded that if I did not fly thither it was impossible to make the journey'. This gift to advertising copywriters was pressed into use to embellish a map. The inverted, left-tilting 'Y' shape of *Imperial*'s 14,000 miles of organised air routes across Western Europe to Cape Town and Karachi were superimposed on an otherwise blank outline of Africa, Europe and Asia. Continent, sub-continent and the places where aircraft landed were labelled in type whose exaggerated twirls conveyed a heraldic air and must have been intended to hark back to a bygone maritime age. A navigational compass completed the illusion. The map legend and Tafur's remark unfurled from parchment-like scrolls. Only sea monsters and puffing cherubs were missing.

The quality of all *Imperial* advertising paled next to that of other big British companies. In the 1930s, for instance, the Dunlop tyre manufacturer placed an extravagant series of full-colour, full-page advertisements in the London illustrated weeklies. In the plushness stakes, the Company sought the upper hand over manufacturers of motor cars, cigarettes, beer and whisky. By comparison, and notwithstanding its connections with Dunlop, *Imperial* appeared to seek nothing at all. The silky car-tyre advertisements played on notions of superior social class but Dunlop did not make the same associations with aircraft tyres for a long time. Then, even before the inaugural flight of the new Ensign-type airliner, Dunlop stole a march on *Imperial* with an advertisement in which a man and woman standing next to a red sports car were shown looking up at the new plane flying overhead. The tyre manufacturer's name printed on the sidewall of a tyre on an out-turned wheel, and on the aircraft tyres, was more conspicuous than the airline lettering.[14]

The cost of advertising, glossy versions especially, was certainly an element behind *Imperial*'s initial reluctance to advertise more frequently and more memorably. The youthful and parastatal airline had other financial priorities. Besides, there were reservations about the appropriateness of airline advertising, brash or otherwise. A snobbish view was that *Imperial* was above advertising. Vulgarity, it was said, was 'the prerogative of cigarette manufacturers, port importers, and others with no pioneer work to do'. Only companies without serious tasks – those who peddled indulgences – could afford to hire electric signs in Piccadilly Circus and plaster London with expensive posters.[15]

When the Cairo–Basra service route opened in 1929, *Imperial*'s Publicity Department published a small, free booklet. 'The New Road to the East' was said to give 'just those natural impressions of the flight which the average passenger would receive, with brief reference to the towns on the route and their historical interest'.[16] An extended, fifty-three-page booklet was issued soon thereafter. It comprised two pieces, most notably a tribute by the veteran aviation journalist, Harry Harper, to 'the world's most wonderful airway'. He described the week-long flight between England and India as 'a romance of super-speed travel'. In a grand panoply, the cover artists wove together London, Mount Vesuvius, the Acropolis, the Nile River, the Great Pyramids at Giza, Baghdad, the Persian Gulf and the Taj Mahal.[17]

Crest

In the mid 1930s, once the Empire air routes and services had acquired better definition, *Imperial*'s advertising increased. The advertising budget for all *Imperial*'s Empire and European services in 1935/36 (approximately £55,000; 4.6 per cent of estimated gross revenue) exceeded that for all publicity prior to February 1936 (approximately £42,000). The vote for the following financial year was raised by 6.5 per cent to £58,000. In both years, a quarter of the expenditure was on advertising in the British press. The next highest categories of expenditure ('general literature' and timetables) accounted for another twenty per cent of the advertising budget in both years. Some fourteen per cent of the budget was directed at general publicity and advertising in the colonial press in each of the four financial years 1935/36 to 1938/39. Over that period, more than half of all such expenditure in Empire flying areas occurred consistently in Southern Africa (average thirty-two per cent) and India and Burma (average twenty-five per cent).[18]

Eye-catching, full-colour publicity posters reinforced advertising hyperbole in the British aeronautical press and in daily newspapers. Expensive poster production became a central element of transformed Empire airline advertising in the second half of the 1930s (3.1 per cent and four per cent in 1935/36 and 1936/37 respectively).[19] Posters were displayed in South African post offices and, in Nairobi, in Torr's Hotel, the grand and fashionable city watering hole owned by Maj. Grogan.[20] Selling the safety of flying was one theme. A 1938 poster recorded that in the previous year 70,000 passengers had flown over six million miles with *Imperial Airways* (the large passenger numbers were due mostly to European air travel). A comforting footnote pointed out that every one of its airliners had four engines. Other poster designs carried repeated visual reminders of the airline's exotic and remote destinations, and nurtured a sunny imperial mentality. The posters also hoped to entice travellers away from the ocean-liners into the clubby atmosphere of *Imperial*'s 'comfort routes'.

The-flight refreshment service was an obvious feature with which to identify attentiveness and comfort. A 1936 *Imperial* advertisement overlaid these themes with that of style. Fougasse, *nom de plume* of the witty and urbane commercial artist C. K. Bird, was commissioned. His posters would have been familiar to users of the London Underground and to devotees of *Punch* magazine that he would later edit. Fougasse's design for *Imperial* showed a cherubic steward trotting confidently above the clouds. He carried a tray with a covered hot dish, and wine. Supported by white wings, the Jeeves-like figure wore a white jacket with a starched collar, and a black bow tie. The poster gave the reassuring impression that in the skies, as on sea and land, the Empire 'maintained its own imperturbable order'.[21] The middle and upper classes were always cosseted and insulated from the rabble. Empire aviation would not change that; on the contrary, it was a service that preserved privileges. These were apparent in the photographs taken of mock-up interiors used during the workshop production of the Empire flying boats: the images were of elegantly dressed passengers (mum, dad, children and granny) posed gracefully beside drinks and daffodils arranged in cut-glass vases.[22] Devoid of any exterior reference points, they have circulated ever since as misleading, myth-making pictures of real *Imperial* air travel.

More realistically, *Imperial Airways* asked Fougasse to portray the relief which flight from London might bring. The first of two juxtaposed line drawings comprising one 1938 advertisement shows a city gent huddling underneath an umbrella in driving rain. He is fumbling in his vast overcoat for coins to pay an impatient taxi driver. A second sketch shows him two days later, carefree, wearing light tropical clothes, riding atop a strolling Indian elephant, shaded from the sun by a vast parasol.[23] The edges of Empire could be reached more quickly than ever; its stressful nerve centre could be fled more rapidly than ever; the pleasure periphery was closer than ever.

One tiny device revolutionised *Imperial Airways'* advertising in the 1930s. For a long time the carrier's publicity lacked a single message and a consistent style. The airline had acquired a diverse fleet of aircraft and could not be associated with one particular type. The airline's name was clear and suggestive enough, but its singularity was undermined by its appearance in advertisements carrying muddled themes and typographies. A stylised acronym might have helped achieve a corporate identity. A corporate badge was an alternative. If the airline ever considered a figuration of Britannia, it was never commissioned. The sole contemporary adaptation was by a British aeronautical tool and sheet-metal manufacturer. Its sexualised advertisements in 1939 fore-fronted a seated, helmeted and bare-breasted young White woman together with lion and Union Jack flag. Text proclaimed that Britannia would rule the skies in addition to the seas.[24]

For the airline, a less corporeal but overtly imperial advertisement might have depicted a network of Empire air routes wrapped over a globe showing dominions and colonies in red. Avoiding such an obvious device (and mindful of the airline's non-imperial air routes in Europe), the designer Theyre Lee-Elliott spent a fortnight perfecting a snappier design that would be associated instantly and uniquely with *Imperial Airways.* He contrived a simple geometrical shape comprising six straight lines crafted in the shape of an arrow or a swallow. The form could also be interpreted as the nose, wings, fuselage and tail of an aeroplane. The blue-coloured logo gestured speed and flight.

'Speedbird', the name given to the new signifier of Empire aviation, appeared on posters, tickets, advertisements, luggage tags, stationery and aircraft fuselages from 1932. The emblem first thrust itself across the Empire courtesy of the philatelic envelopes issued by *Imperial* to mark the start of its Indian, African and Australian air mail services. The airline sent its own officials special greetings covers overprinted with route maps, dates, the Company's name and 'Speedbird'.[25]

As shown by the display of paper items in the commercial printing section at the Royal Academy's 1935 exhibition of British industrial art, the elegantly simple and flexible 'Speedbird' motif could be manipulated with ease in pattern repeats, and in association with words and images to generate suggestive illusions. At a stroke, *Imperial Airways* advertising shifted from being merely informative to being memorable and visually seductive. The modernistic, sleek form imparted an élan surpassing the photographs of lumbering aircraft. In *Air, Imperial*'s Sports and Social Club magazine, Lee-Elliott wrote that he submitted his design with the pious but confident hope that although 'Speedbird' was a bit stiff-necked for a goose, it would lay a good number of golden eggs.[26]

Everywhere it was seen the symbol of Britain's newest imperial transport announced Empire and proclaimed aviation. In its early days, *Imperial Airways* advertised rather laboriously and literally. Now it could do so with more flair, including on the road coaches used to transfer passengers to and from Croydon and, later, on the Southern Railway connecting London to the flying boat base at Southampton. Like the green morocco leather furnishing in the silver flying boats, the landside service was the epitome of England. Sir Harold Nicolson relished the homecoming when returning to London in 1937 in a special Pullman marked 'Imperial Airways – Empire Service'. After disembarking from *Centaurus* at Southampton, his train journey on a snowy Sunday afternoon reconnected him with the cultural markers of Sunday newspaper, tea and hot butter toast, and Craven A cigarettes. 'HOME', he wrote emphatically in his diary.[27] Up and down the railway line, at countryside station platforms, at level crossings and from behind hedgerows, trains propelled the Empire airline into view. Nicolson, like many who used the flying boats, was not photographed: either *Imperial* had decided there was no more publicity to be squeezed from photographing eminent passengers, or Southampton was beyond the focal length of London pressmen.

In the 1930s, the commercial artists who worked on the *Imperial Airways* account – including employees of the advertising agency which handled the publicity for the elite grocers Fortnum and Mason[28] – reminded millions of Britons that 'their' airline operated the world's longest air route. Slick graphics and smart copy concealed the crumbling cohesion of the Empire. 'Speedbird' appeared on posters advertising *Imperial Airways* in French, Arabic and Hebrew, as well as English. As time passed, publicity was aimed at a broader clientele than celebrity and novelty passengers. Civil servants were reminded that flying meant longer home leave in England. Serving officers were courted with the promise of fare rebates. A 1937 poster fastened patriotically

Figure 12 Speedbirds propelled by steam. *Imperial Airways* London–Southampton train at the rear of *Imperial Airways* House, Victoria, London, 6 June 1939.

onto the Coronation of King George VI ('The World's Greatest Event'), and emblazoned a regal theme with text suggesting that visitors use the time saved by air travel to prolong their stay in England. Another version of the poster listed a series of summer events in England, beginning with the Ideal Home Exhibition and ending with the Derby.[29]

'Australians have travelled in their hundreds by aeroplane for the Coronation', readers of the paper *Great Britain and the East* were told.[30] Those who could not fly to London were served by *Imperial Airways* in another way. News reports and film footage of the Coronation would have been flown east to reach expatriates and the King's other subjects in India, Australia and British Africa. Westward, across the Atlantic, images of the royal event would reach Canadians and 'the former subjects of King George the Third' in the United States via Europe. Without any apparent misgivings about dependence on German technology, three aeroplanes loaded with photographs and newsreels would dash from London to Frankfurt from where the German airship named after President Hindenburg would take the material on the second of the 1937 season's regular flights to New York. Its scheduled departure date had reportedly been postponed to accommodate the consignment.[31] In a quirky reverse to the delayed receipt of news about the declaration of American colonial independence in 1776, films of the Coronation did not reach the United States quickly. This time the transmission delay was because the *Hindenburg*, crown jewel of German aviation, flagged in a swastika, crashed six days before the Coronation. There was no British air service to fly the films to New York, or to fly stranded airship passengers to London for the Coronation.

Exhibitions

The British Empire airline did not confine its advertising to paper. A nine-foot-long window display designed to show the speed advantage *Imperial Airways* had over road and rail transport worked with a design of four distinctive wheel types depicting the evolution of overland transport. Each successively larger wheel was labelled according to the transport chronology with which it was associated. Surmounting all the wheels was a drawing of an aeroplane fuselage to which the label '100 miles per hour' was appended. Seen quickly, the image had militarist overtones. The front end of the aeroplane appeared to emerge like a bullet from a cannon mounted on a four-axle gun carriage. If he spotted the contradiction with civilian flying, Sir Philip Sassoon remarked only that the display was of a national asset and demonstrated aspects of the 'mighty Empire of which the public was privileged to be citizens'. Rather more ambitiously, an art critic declared that the exhibition captured the enlarged experience afforded by air travel. In an indulgent aside, the critic suggested that flight made reality less perplexing by exposing its cosmic order and by merging history and geography.[32]

In the autumn of 1934, *Imperial Airways* and *The Times* banded together to mount an exhibition entitled 'Flying over the Empire'. The venue was the gallery at Gieves, the venerable men's outfitter in London's fashionable Old Bond Street. The show comprised a large folding screen on which was mounted a map of Empire airways, models of *Imperial* aircraft and photographs depicting imperial scenes. One of several dioramas featured an outsize airliner outside the Sharjah rest house. Aerial photographs of the Sudd, locust swarms and Bor elephants were unlike anything most shoppers would have seen before. A modernist photograph of inanimate aircraft engines, themselves objects of wonder, was entitled 'Horse Power'. The exhibition toured Britain at the end of its ten-day London showing. Four additional exhibition sets were prepared for display in provincial towns. Empire aviation was being exposed beyond the closed world of London hotels, clubs and societies. The residents of Bolton, Burton-on-Trent, Buxton, Derby, Folkestone,

Gateshead, Halifax, Lincoln, Plymouth, Scarborough, Sunderland and Tonbridge were snared in the imperial net.[33]

The booklet-cum-souvenir which *Imperial* issued to accompany the exhibition pitched its enterprise on a lofty plane, partly by decorating it with quotations from two prominent British men. They lent noble pedigree and purpose. First, Dr Samuel Johnson's eighteenth-century philosophical romance 'Rasselas' included a dissertation on the art of flying which yielded this snippet: 'I have long been of the opinion that instead of the tardy conveyance of ships and chariots, man might use the swifter migration of wings.' Johnson's addendum that 'the fields of the air' were 'open to knowledge' was the kind of teaser in which propagandists delight. Secondly, the archetypal lone hero, Lawrence of Arabia, then an RAF airman, was quoted as writing in his last letter to his poet friend Robert Graves that 'the conquest of the last element, the air, seems the only major task of our generation'.[34]

In 1935/36 the 'Flying over the Empire' exhibition was remounted at London's Science Museum for a two-month Christmas showing. Renamed 'Empire's Airway', admission was free. Exhibits followed in Ottawa and at the Canadian National Exhibition in Toronto, and in Palestine. The New Zealand capital (Wellington) and the South African commercial capital (Johannesburg) are mentioned in the records; 400 people visited a showing at the Transvaal Museum in Pretoria.[35] During an Australian tour, the British High Commissioner in Australia, Sir G. Whiskard, opened the 'Empire Air Routes' photograph exhibition in Sydney in May 1936. The show also formed part of Adelaide's Centennial Exhibition.[36]

When the exhibition returned to Charing Cross in mid 1936, it had acquired a walk-through, cross-section cutaway of aircraft interiors. The information presented to visitors (who, in a private preview, included 'many hundreds of distinguished people') included the claim that each day *Imperial* and its associated overseas companies flew a distance greater than three-quarters of the earth's equatorial circumference. The hold over the globe was spiritual too, as an attribution to le Corbusier intended to show: 'THE AEROPLANE IN THE SKY CARRIES US ABOVE MEDIOCRE THINGS'. The architect of modernism actually used the words 'our hearts' rather than 'us', and they capture better his view that aircraft were not merely utilitarian but were symbolic of a new age, diviners of a romantic 'machine civilisation'. His own paean to the plane accompanied a photographic exhibition in 1935. It incorporated several striking close-ups of exterior surfaces and shapes of *Imperial* aircraft; some images were clipped from the 1932 film 'Aero Engine'. The more prosaic slogans at Charing Cross indicated merely that 'THE SPEED BIRD OF IMPERIAL AIRWAYS IS

THE SYMBOL OF THE NEW AGE' and that 'THE SUN NEVER SETS ON THE ACTIVITIES OF IMPERIAL AIRWAYS'.[37] The jingle writers failed to exploit Edmund Burke's 160-year-old aphorism that 'a great Empire and little minds go ill together'.

The *Imperial Airways* photographic exhibition in Raffles Hotel, Singapore, was opened in September 1937 by Governor Sir Shenton Thomas, himself an *Imperial* passenger from Singapore to London in April 1936.[38] By the time *Imperial Airways* exhibited again at the Canadian National Exhibition in 1937 (audience 300,000), more than a million people had seen the maps, models and other displays. In Britain, the airline continued to tap and create public curiosity by arranging a three-day exhibit in a train carriage at London Victoria terminus in July. Like France's 1936 aeronautical road show ('Exposition Circulaire de Propagande Aeronautique'), the *Imperial Airways* version was wheeled through the country: Blackburn, Glasgow, Edinburgh, Newcastle, Sunderland, Middlesborough, Harrogate, York, Hull, Leeds and Bradford. A second cycle took the exhibition train through West England and Wales. Stops were made at Portsmouth, Bournemouth, Exeter, Taunton, Bristol, Cardiff, Swansea and Gloucester. Most visits lasted three days, the first being devoted to invitation inspections by local business firms. The public had entry to the exhibition on the other two days. Maps, charts, aircraft models and radio transmitters showed off the comfort, safety and reach of air travel. By the time the exhibition train had completed its rounds early in 1938, *Imperial Airways* would have hoped that the 135,000 visitors had a new regard for and pride in flying imperially.[39] Eight per cent of advertising budget was spent in 1937/38 on the 'Empire's Airway' exhibition train.[40]

Alongside high-profile exhibitions, *Imperial* also undertook more prosaic publicity in an effort to make flying part of British culture. Sometime in 1932 or 1933, a Mrs Doris Moore was appointed as *Imperial*'s first liaison and sales representative for England and Scotland. A rare account of her work (in an Australian newspaper) reports her having contacted 1,000 travel agencies throughout Great Britain. Her car trips averaged 11,000 miles a year. In addition to meeting agents, for six years Mrs Moore lectured at schools, societies and clubs on aspects of air travel, showed films and held exhibitions, and displayed a variety of models of aeroplanes and photographs.[41] From 1934, she would have had access to a collection of *Imperial*'s portable lantern slides. Anyone could borrow this pre-set programme of transparent images free of charge from the airline for public lectures. The first of these was called 'The Empire's Airway' and 'The Future is in the Air'. The slides were used in more than 400 lectures given in Britain in 1935/36 by people not attached to *Imperial*.[42]

In 1936 *Imperial* issued a revised version titled 'Wings over Empire'. The new pack included 100 slides and a companion booklet with text for reading out as each slide appeared on screen. As on fifteen picture postcards dealing with *Imperial*'s African airway, the usual array of images was recycled: views of aircraft, airports, ancient buildings, landscapes, wild animals and people along the airways followed one another relentlessly. The slide-show scripting was informative but bland; viewers and listeners got a sense of a grand enterprise rather than a problematic one. A black-and-white text matched the monochrome images. Only the closing words flirted with colour: given its youth, the results of British air transport were astonishing; future developments would inspire pride in the Empire airway and give hope that its peaceful ways would extend happiness and the blessings of civilisation to all people. This spoken finale probably ended the talks given by three lecturers engaged by *Imperial* in the winter of 1937/38. Their next 143 presentations on 'The Empire's Link' used a new set of slides. Audiences ranged from the fifteen hardy women who turned out as members of the Tooting townswomen's guild, to the 500 people who thronged the lecture given to the Liverpool Chamber of Trade in the city hall.[43] Seventy slide shows helped to make audiences in Sweden and Denmark conscious of *Imperial Airways* and to counter-balance American and German airline advertising.[44]

Subliminal shows

A lot of *Imperial Airways* advertising was on its own account, but the airline and the imperial spirit benefited by a considerable amount of free publicity. This was embedded in technical and general-interest magazines, journals of learned societies, news reports and photographs, press statements, published letters, and newspaper summaries of corporate accounts and shareholder meetings. Even model toys made reference to the Empire airline. From 1934, the 'Skybird' range of commercially produced model airplane kits included *Imperial* aircraft with decal registration letters appropriate to either the Cape or India services.[45] The Dinky toy manufacturer of Liverpool brought out its own *Imperial Airways* range in the mid 1930s. The airline stood to gain too from the propaganda work done by the Air League of the British Empire, including displays at the Schoolboys Exhibition in London in 1936 and a programme of lectures, speakers and slides on 'Linking the Empire by Air'.[46]

An additional source of indirect *Imperial Airways* advertisement was given by commercial organisations seeking an association with the Empire and the Empire airline. The technological modernity, speed and romance of scheduled (as opposed to dare-devil) flight were eminently exploitable. In the commercial press, for example, *Imperial* benefited from advertising by a spark plug manufacturer. These invisible items of such little intrinsic value acquired a reputation only in association with a successful engine. Aircraft engines beating faultlessly across thousands of Empire route miles in all kinds of climates were heaven sent. A full-page picture of an *Imperial* aircraft superimposed on a map of the air route across Europe and onward to Karachi and Cape Town soaked up patriotism and imperial identifications. A distributor of 'Aero' wristwatches boasted simply that both *Imperial* and the de Havilland Aircraft Company had 'proved' its product. Carr's biscuit manufacturer, once contracted to *Imperial*, advertised this association proudly.

One Shell Oil advertisement grabbed attention with headline text comprising just the two words 'Imperial Airways'. A second line referred to 'The World's Finest Air Service'. On a third line, in smaller type, it was explained that the airline's qualities testified to the qualities of the oil. Appearing in the pages of the *Sphere* while civil flying was still a rarity, and before the Empire service had begun, the advertisement was actually targeted at wealthy car owners, and reinforced associations between class and opportunity. In 1932 the work of the Empire airline was advertised as the benchmark of lubricant efficacy. Text pointed out that eight return journeys to the moon would be less than the four million miles flown by *Imperial* using (British) Wakefield Castrol. At least once, the Short Company that built *Imperial*'s second-generation flying boats placed a double-page advertisement trumpeting the airline's service. In 1938, the annual civil air transport edition of the *Aeroplane* carried a two-tone photograph of the new Empire-class aircraft, a world route map and text reading 'Short Flying Boats – the Vital Link in Empire Air Communication'.[47]

Commercial organisations did not only use the Empire airline as a prop for their products. Another popular strategy was to embed product advertisement in a message of congratulation to private pilots who had just set some new flying record. Conspicuous individual achievement was made to resonate with British national prowess but rarely with imperialism; many readers would have elided the two. One fairly subtle exception occurs in a cartoon-like advertisement placed by Shell in periodicals such as the *Aeroplane* and the *Spectator* in 1933. Although there was no explicit mention of Empire, or of imperial aviation, the advertisement draws on and reinforces imperial archetypes even while apparently undermining them. The humour relies on subverting artistic juxtaposition, itself a classic device for showing cultural contrast and depicting relations of subservience and dominance. A couple of

Africans squat next to their food bowls placed on the ground in a forest clearing. The artist spared the caricature cooking pot, but not the loincloths, amulets, bangles or ear-lobe discs. A pilot sporting superfluous flying goggles and cap stands nearby; he has the knee breeches, boots, tanned colouring and extrovert bearing of a Frenchman or Italian rather than a Briton. Trifling detail would not compromise the message. The pilot points excitedly to his enclosed, German-looking monoplane in the background. The plane has flown from the known world to a mock place. Its fuselage is labelled 'Non-stop London to Juba-Juba'. Unusually, the disinterested Africans have their comeuppance. One of them pre-empts the pilot: 'Yes I know you did it on Shell – go away'.[48] In colonial circles it was more customary to poke fun in the other direction, as in a cartoon showing an African chasing an aircraft shadow in an effort to remain in cool shade.[49]

In the category of *Imperial Airways* 'advertorial', news items about Empire air mail rate high. Photographs of mails being handled at an airport pictured modernity, haste and linkage. In December 1931, cinemagoers could see newsreel images of Kingsford Smith at Croydon where he landed Christmas mail from Australia. A photograph in the *Sphere* showed him in his flying gear standing alongside workers at the *Imperial* mail handcart on the apron. The picture implied he was working for (not rescuing) the British carrier, the very one which was to throttle his ambition to operate his own airline on Empire routes. The following year, the photograph on the front cover of the Christmas-eve issue showed men loading the Africa-bound mails. The caption mentioned horsepower replacing Santa Claus' reindeer, and referred to 'freight of sentiment'. The paper cooed that nothing epitomised the giant strides of twentieth-century transport more than air mail enterprise. Unintentional associations sometimes crept unnoticed into photographs. One in the *Sphere* proudly showed men unloading the first mails to arrive by air from Brisbane, on Christmas eve 1934. Unpatriotic stage managing (and picture cropping) meant that air mail which arrived aboard an *Imperial* aircraft was shown being trundled away in a cart labelled 'Air France'.[50]

Cigarette cards were another class of promotional device. From 1936, smokers and collectors – a large set across all social classes and regions of Britain and Empire – encountered air imperialism courtesy of Lambert & Butler and John Player's, brands in the Imperial Tobacco Company stable. Several full-colour cards in a series of fifty inserted in packets of Waverley and Rhodian cigarettes depicted maps, airliners and important stops *en route* to Africa and Australia. In 1938 the Senior Service brand of cigarettes was sold with forty eight cards in a series on flying. The set included a picture of the 'parliamentary visit' by the Short Calcutta flying boat, one of the seven Handley Page-type 'sister aircraft', and one of an 'Empire Class' flying boat and its passenger cabins. A set of fifty Churchman cigarette cards issued in 1939 and titled 'Kings of Speed' included one of *Imperial*'s Capt. G. J. Powell, who had piloted the flying boat *Cambria* on a record run,[51] and one of Capt. D. C. T. Bennett, 'Senior Master of *Imperial Airways*', who was prominently identified as Australian. One of the 1936 'Pictures of Britain' cigarette cards series issued by the J. A. Patreiouex company featured Croydon aerodrome terminal. Issued by Stephen Mitchell & Son in 1937, the set 'Our Empire' included *Imperial*'s flying boat *Caledonia*.

A final category of incidental airline advertising occurs in commercial art. Three examples show that the monopoly which *Imperial* enjoyed as Britain's sole carrier on overseas air routes was underlined in illustrations of the newest and biggest civil aircraft and airports in Britain. Planes, names, routes: the most modern transport technology and the assertion of Empire went hand in hand. For instance, the 1933 annual flying number of the *Bystander* – a 'society and fashion' magazine – had a full-colour cover showing passengers embarking on a passenger aircraft marked as one owned and operated by *Imperial*. A year previously, a reproduction painting of flying activity at Kisumu depicted aviation servicing the Empire.[52]

A more opaque and more gallant imperial reference was in a dark front-cover illustration of the 1930 *Wonder Book of Aircraft*. The front section of an unmistakably British bi-plane is shown circling an airport (Croydon, presumably) and a city whose lights twinkle beneath like stars in an inversion of earth and sky. The impressive length and wingspan of the aircraft is left to the imagination. Homeward bound, apparently, the plane is drawn to the beam from an airport beacon. But the aircraft is not a tiny, helpless moth; it is a behemoth, and it is under careful supervision. In an open cockpit, the capped head of the minute figure of a pilot is seen peering through the windshield. He steers his machine and his passengers heroically. A silent row of figures sits alongside the square windows in an enclosed cabin lit a glowing gold colour. In the night sky, the lower aircraft wing shadows some of the fuselage from the airport beacon, but in the instant that the airborne painter had to capture a fleeting coincidence of light and lettering, the golden beam picks out the last thirteen letters in the name *Imperial Airways*.[53]

Airport and aircraft

Direct and indirect airline advertising shouldered only some of the burden of dressing and parading imperial flight. Aeronautical hardware assisted. Notably, aircraft and the ground facilities contributed to the image and spectacle of an ongoing and progressive Empire. An early marketing coup was arranged by the Air League of the British Empire: in 1928 *Imperial* moored its new flying boat in the Thames just twenty yards from the House of Commons terrace. Parliamentarians, civic officials and Lloyds insurers were all invited to inspect the new tool and totem of Empire.[54]

Compared to their successors, aeronautics and aviation in the 1920s and 1930s were feeble. In their day, however, the sights were awesome, the sensations elemental. At air stations, incarnations of latent power were paused and poised to launch. Airports were sites of public awe tinged by visceral terror, not of fire, steam and smoke, but of vertigo, speed and shrill noise. Aerodromes were places where practice rather than comprehension integrated a new generation of machines into modern industrial society. Airfields were a locale where technology constructed new and powerful social experiences, and invigorated national pride. A British inflection of 'the technological sublime'[55] confirmed the continuity of Empire, and paraded the engineering feats that updated, justified and entrenched it.

Croydon aerodrome was the homeport and anchor of the newest overseas transport services. It was the heart of an imperial embodiment in which airways were new arteries. The place had the frenetic buzz appropriate to its status as Britain's pre-eminent metropolitan airport. As early as 1926, the case was told of one ten-year-old boy who haunted the airfield: 'his friends are the mechanics, the pilots are his Drakes and Nelsons'. A pilot at Croydon implored the writer 'for goodness sake to tell people what the air is going to mean to England'. His wish was honoured in the same unsigned article, which asked the 'crabbed counsel' of cynical adults not to dampen the aviation ardour of British youth: 'the public enclosure at Croydon is crowded with children; in their bright eyes, in their sane minds, live the hopes of conquests wider and worthier than any our race has yet made.'[56]

A visitor in 1931 also sensed the imperial atmosphere of Croydon. Even before the start of regular Empire air services, an inert direction signpost anticipated the gateway status of the aerodrome: its arms showed the way to Cairo (2,000 miles), Karachi (4,000 miles), Johannesburg (6,000 miles) and Sydney (11,000 miles). Porters swarmed, propellers whirled. The to-and-fro travel of the *Imperial Airways* motor coach, the quick pass through the bustling booking and waiting hall, and the rush through the customs area to a departing plane, left an indelible impression of managed intensity. On the ground, if not in the air, it seemed to one writer that *Imperial*'s motto should be 'speed casteth out all fear'. Aircraft were said to arrive and depart the gargantuan lungs of Croydon 'with a pyrotechnic force and a disconcerting punctuality'. The palpable energy – the dazzle and din of Empire – was likened to 'the materialisation of an aerial Bradshaw [railway timetable] set to the murderous, the annihilating Z-Z-Z-Rugh-Z-Z-Z-Rugh of mammoth engines purring, as volcanic eruptions might purr'.[57]

Host to this supernatural scene, the airport complex was compelling by virtue of its sheer size. The grass landing field covered 300 acres. Excepting for their scale and number, the aircraft hangars were not a unique airside facility, but *Imperial*'s offices, the nearby Aerodrome Hotel and the Air Ministry's administration building were exceptional. The world's first custom-built civil airport symbolised everything that was exciting, modern and forward looking about overseas air travel. The 1935 edition of *Every Boy's Hobby Annual* had no doubts about the importance of a terminal 'at the end of a vast system of air routes extending over the whole civilised world'.[58] At the heart of the administration building was an art deco booking and waiting hall. The *Imperial Airways* freight- and passenger-handling section occupied a conspicuous position. Like a proscenium arch on the imperial stage, enormous doors at the end of the hall led onto the apron where nose-in-the-air aircraft haughtily awaited boarding. Emerging from the solidly reassuring architecture of stone building and clock-faced tower – echoes of municipal library and country church – a stroll of twenty paces put the Empire at one's feet.[59] Grandeur was less conspicuous at the edges of Empire.

Befitting and enhancing its status, Croydon aerodrome was one of the focal points of celebrations marking the silver jubilee of King George V and Queen Mary in May 1935. The enduring institutions of monarchy and marriage merged with modernity. The borough newspaper crooned about the worldwide fame of Croydon airport seven years after the terminal was opened by Lady Hoare. Ever an accessory to aviation, she did so shortly after her husband initialled the agreement between *Imperial Airways* and the British Government for a service between London and India. Tactfully omitting to mention its location on the so-called Purley Way, the paper noted proudly that the landing field had become a gateway to (earthly) Empire. A former *Imperial* employee recalled that people all over the world had heard of Croydon: 'it represented the last word in international aviation – everyone in Britain was proud of it'.[60] Not so, one imagines, the water-men, lighter-men and inn-keepers who had made a living at the Thames docks around which the seats of imperial authority and the offices and warehouses of Empire shipping and merchandising had clustered for centuries. Their affections would not have transferred easily ten miles south of London.

In the late 1930s Croydon the usurper itself lost its imperial role.

Figure 13 Imperial periphery: Empire flying boat terminus, Durban (South Africa), 1937.

When the flying boat boarding pontoons were installed at Hythe (Philip Sassoon's constituency) close to the Southampton mail boat base, Empire shipping and aviation were brought closer together again. Viewed from distant Plymouth, it must have seemed that history had been doubly traduced. That port town had famous associations with the New World and with British maritime enterprise. In 1914 Shackleton had set sail from Plymouth on a journey he hoped would open a trans-Pacific imperial link. Even in its failure his journey became a symbol of British endurance. An imperial flight had also originated at Plymouth. On a sultry dawn in September 1928, Sir Philip Sassoon flew away on his expedition to the Middle East and India. His flying boat, 'a great, grey, graceful shape riding lightly on the leaden water', seemed to him the embodiment of power and speed. The sight of the aircraft triggered a tingle of excitement. Another emotion registered after he took off. Through his cockpit window he gazed down on the statue of Sir Francis Drake on the Hoe, and sensed 'the inspiration that it must ever have for all our people'. That stimulus and reminder never transferred to Southampton. The liners of the air confronted exclusionary maritime sentiments: 'the ships were there first, and the two don't go together'.[61]

More than its bases at Croydon and Hythe, *Imperial*'s aircraft fleet was part of the iconography of Empire, and an expression of patriotism. The very name of the airline helped resurrect imperial discourse, keeping the notion of Empire in everyday use and consciousness. In addition, aircraft and their crew were meant to revive the swagger of imperialism. The Royal Air Ensign fluttering from the cockpits of Royal Mail Aircraft signalled a British presence at airports around the Empire. Onlookers saw that Britain was indeed great. Sir Harry Brittain's companion noted the importance of conspicuous flagging when on his 1934 trans-Africa flight. There was a time, he wrote, when British people were proud of the wind-whipped flag of the mercantile marine in out-of-the-way parts of the world. In the air age the sky-blue *Imperial* pennant thrilled British citizens in the Empire outposts.[62]

The names of some *Imperial Airways* aircraft were themselves emblematic. The criteria for selecting the names of Empire cities to paint on the sides of aircraft are unclear. London was an obvious choice, but not Ottawa, which was not (and never would be) on the *Imperial Airways* route map. Inexplicably, *City of Salonica* became *City of Swanage* and *City of Wellington* became *City of Arundel*. A commentator remarked that *Imperial* did have a habit of digging up cities that nobody had heard of before. The African names referred to well-known cities at the northern and southern extremities of the continental airway: Cairo, Alexandria, Khartoum, Pretoria, Cape Town. Intermediate stops along the Middle East–India route were better represented: Jerusalem, Baghdad, Teheran, Basra, Karachi, Jodhpur, Delhi. No cities further east were commemorated in aircraft names excepting Melbourne and Wellington.

As tokens of Empire, aircraft names had to bear simmering tensions between its core and edge. A nationalistic and conservative Afrikaans-language Pretoria newspaper jibbed at 'idiotic names on aeroplanes' in 1932. The target was the 'unpronounceable coolie names' of *Imperial* aircraft flying in South African skies: *City of Baghdad*, *City of Delhi* and *City of Karachi*. C. G. Grey bristled. Inverting the argument, he suggested renaming the *City of Pretoria* that Cobham was then flying in England. He thought that British joy riders who objected to 'kaffir names' would prefer to fly in, say, *Borough of Walthamstow*.[63] The derivation of 'Pretoria' happened to be Afrikaans rather than African, but etymological error counted for little in a spat.

Naming flying boats *Hengist* (stallion) and *Horsa* (horse) was an odd imperial reverse if it was done mindful of the brothers who were reputed to have led Germanic forces in Britain in Anglo Saxon times, and who founded the Kingdom of Kent and its royal house. Possibly, the classically trained leaders of public affairs in Britain between the wars were honouring horse deities whose alliterative names recalled the founding legends of the Indo-European world such as Romulus and Remus. Most aircraft in the *Imperial* fleet bore less obscure names, even though their imperial resonance was indirect. At least the classical names given to

Imperial Airways aircraft had meanings that did not contradict imperialistic thought. Names like *Achilles, Apollo, Avatar, Challenger, Coriolanus, Hannibal, Hadrian, Helena* and *Hercules* (chosen by airline directors from entries in a *Meccano Magazine* competition) alluded to power, authority and stature. Some fuselage badges featured fierce Greco-Roman artwork: an armoured gladiator, a winged horse, an archer, a horseman, a galleon. A sentry wearing tartan trousers was less combative. The dandy in high boots, cape and feather hat might have been a Drake.[64] *Andromeda* and *Centaurus* had mythical and heavenly connotations. Like the grand image inscribed in the name of the Empire airline itself, these names were rather wishful descriptors of lumbering aircraft whose speed and altitude were anything but stellar. Image and reality were easily confused when packaged enigmatically in vague allusions.

Only a few aircraft were given names to carry forward a sense of the Empire's maritime past. *Cabot* happened not to be an English navigator and *Corsair* had unfortunate piratical associations, but *Frobisher* had been a renowned Elizabethan sailor. The link between Empire and sea was barely sustained either during brief aircraft-naming ceremonies. These did not copy ship-christening procedures. When Lady Hoare drew a silk chord to unveil the names of five new aircraft in 1926, she did not smash a bottle against their sides. But champagne was almost certainly drunk at the ensuing luncheon. By then, guests would have been bubbling with pleasure at seeing a miniature Union Jack flag fluttering from the rear struts of each aircraft, and watching homing pigeons released from the cockpit of the first aircraft to be named.[65]

The flying boat named *Golden Hind* evoked a glorious seafaring lineage like KLM's *Flying Dutchman*. Anticipating a different future for the aircraft that was due to enter service as the Second World War broke out, firms connected with its construction made a great play on its Drakeian ancestry in their advertising. None of the flying boats that flew *Imperial Airways* colours had individual names with nautical associations. There was no need, for the type was self-referencing. The riveted metal hulls of the first-generation ('Calcutta') flying boats distinguished them from landplanes and, like the water on which they rested, justified their generic name. Inside the slightly larger twenty-four-seat second-generation ('Empire') flying boats, sections were dubbed the 'bridge' and the 'promenade deck'. When these flying boats departed Southampton on through-services (no need to change aircraft), they were spoken of as vessels departing on voyages from their moorings. The label 'Empire' may have been no more than the most obvious alternative to an alphanumeric engineering designation by the Short aircraft manufacturing company for aircraft ordered for the

Empire air mail services. Yet the label 'Empire-class' was more than a technical classification; it was also an imperial insignia. 'Class' meant category, but its classy connotations were those of grandeur, comfort and impeccable high taste.

More than any landplanes, *Imperial*'s flying boats helped to prolong Britain's seafaring tradition. These airliners were the bluff that the age of maritime Empire was not past. It had simply become aero-nautical. By virtue of their generic name too, airships would have carried a maritime charge, whether singly or under the collective noun 'fleet' applied to all transport vehicles. The airships *R100* and *R101* might have struck a louder imperial chord if they had been given names more apposite than numerals on a draughtsman's specification sheet. In Parliament in 1929 Sir Harry Brittain asked Hoare to follow nautical custom and name the airships more imaginatively. Suggestions published in the correspondence columns of *The Times* included *Queen of England*, *Flying Briton*, *The Great White Whale* and cutely, from Shakespeare, *Oberon* and *Titania*.[66]

Once airships were scrapped, landplanes became the chief inscriptions of winged Empire. But it was flying boats which evoked a sense of continuity with the past that made the new order less radical and more intelligible. British people, it seemed, could never escape the sea: 'to overcome the difficulties of their isolated geographical position, they must go under or upon or over the oceans of the world'.[67] In that context the flying boats roused a deep nostalgia. In 1936 the *Manchester Guardian*'s aeronautical correspondent watched *Canopus* roll down the slipway at Rochester. The sight was stirring, even in an age when air-minded people might have been expected to tingle more at the sight of a new airliner than they had at the recent launch of the ocean-going liner *Queen Mary*. The flying boat launch did trigger a frisson: would it fly to Sydney or Durban; would a night flight across India wake ryots in their mud huts and scare jackals from their prey; would it soar over wilderness discovered by Livingstone; would it float where Cleopatra's barge anchored?[68]

Canopus was certainly meant to land near the ancient Egyptian port town of the same name. The makers and operators of the flying boat surely would have wanted it to be a bright star like its astronomical namesake. Happily, the new aircraft did not copy the *Titanic* and the *R101* by crashing on its maiden commercial outing. The jinx was never completely broken, however: air steward Jefcoate was killed when *Capricornus* crashed on its maiden commercial flight in March 1937. Even *Canopus* had a narrow miss. It was saved from fire when anchored some distance from the Lake Bracciano shoreline on one of its early flights. Once the fire extinguishers were exhausted, a human chain passed buckets of water out of the aircraft galley. Finally, green pea soup (which was to have been served as Potage St Germaine) doused the flames. Other incidents ended less happily. The list of flying boat mishaps became a poor advertisement for British technology round the Empire. Of the twenty-nine flying boats *Imperial* ordered, eight had crashed and one had burned out by mid 1939. In unfortunate timing that summer, *Vogue*, the London fashion magazine, published a half-page *Imperial* advertisement soliciting flying boat custom two days after *Centurion* capsized on the Hooghly River at Calcutta (four passengers and five crew were saved) and five days before *Connemara* burned to destruction at Hythe.[69]

Notes

1 *Bystander* (29 April 1936), p. 219.
2 *Air Travel and Commercial Air Transport,* 1 (1934), p. 40.
3 Founded in April 1924, the same month that the British Empire Exhibition opened at Wembley, *Imperial Airways* did not show itself there. But inattention at Wembley to Empire air routes is surprising as they had been discussed in some detail since 1917.
4 *Daily Express* (18 September 1929), p. 8. Poster reproduced in *Commercial Art*, 14 (1933), p. 80, and in H. S. Menzies, *All Ways by Airways* (London, Imperial Airways, n. d.).
5 Cluett, et al. (eds), *Croydon*; RAFM, B 343.
6 BL, WP 15116; Cluett, et al. (eds), *Croydon*, p. 72; *Imperial Airways Gazette* (September 1934).
7 *Field* (21 May 1932), p. 14; BL, WP 15116.
8 BAAM, AW/1/7328; PRO, CO 323/1616/2; *Imperial Airways Gazette* (June 1936).
9 For extended discussion see C. D. Bhimull, 'Empire in the Air: Speed, Perception, and Airline Travel in the Atlantic World' (PhD thesis, University of Michigan, 2007), pp. 73–122.
10 *Imperial Airways Gazette* (April 1933; April, July, August, September, October 1934; January, April 1935).
11 H. Burchall, 'The political aspect of commercial air routes', *Journal of the Royal Central Asian Society*, 20 (1933), 70–90; 'British commercial air routes,' *Scottish Geographical Magazine*, 49 (1933) 193–213; 'The politics of international air routes', *International Affairs*, 14 (1935), 89–107.
12 *The Times* (5, 11, 17 July 1929); (5 November 1934), p. 11; (12 September 1935; 17 July 1936); *Airways* (August 1929).
13 Sims, *Adventurous*, p. 54.
14 *Sphere* (7 July 1938), p. 75; (30 July 1938), p. 178.
15 *Flight* (2 March 1933), p. 202.
16 *Flight* (11 April 1929), p. 306.
17 Thanks to Daniel Kusrow for this detail.
18 BAAM, AW/24/216, *Imperial Airways* Board Papers, 17 March 1936; AW/28/254: Memorandum to the Board, 17 May 1938.
19 AW/24/216, *Imperial Airways* Board Papers, 17 March 1936.
20 *Imperial Airways Gazette* (December 1932).
21 J. Morris, *Riding the Skies: Classic Posters from the Golden Age of Flying* (London, 1989), p. 9.
22 Cassidy, *Flying*, p. 48.
23 *Air Review* (April 1937), p. 36; *Illustrated London News* (18 June 1938), p. 1140.

24 *Aeroplane* (26 July, 2, 16 August 1939).
25 N. C. Baldwin, *Imperial Airways* (Sutton Coldfield, 1950); *Imperial Airways Gazette* (July 1934; August and October 1936); *Air Travel and Commercial Air Transport*, 1 (1934).
26 *Imperial Airways Gazette* (February 1935); *Air* (April 1938), p. 8; D. Scott, 'Air France's Hippocampe and BOAC's Speedbird: the semiotic status of logos', *French Cultural Studies*, 4 (1993), 107–27.
27 Balliol College Library, Sir Harold Nicolson Diaries, 6 and 7 March 1937.
28 *Commercial Art*, 14 (1933), 75–60; 19 (1935), 190–1.
29 Morris, *Riding*, pp. 40–1.
30 G. Bolton, 'Airships, aeroplanes and the Empire', *Great Britain and the East*, 48 (1937), 685.
31 *The Times* (30 April 1937), p. 11.
32 *Commercial Art*, 14 (1933), 78.
33 *Imperial Airways Gazette* (August, September 1934); *The Times* (10, 12, 17, 18, 25 July, 2 August 1934).
34 Imperial Airways, *The Empire Flying Boat*.
35 BAAM, AW/24/216: Board Papers 17 March 1936; *Imperial Airways Staff News* (14 July 1936); *Evening Post* (Wellington) (26 September 1935), p. 25.
36 *Sydney Morning Herald* (26 May 1936); *Imperial Airways Gazette* (June 1936). Between 1935 and 1937 *Imperial* commissioned László Moholy-Nagy, in exile in London from the famed Bauhaus, to produce modernist publicity material, including an Empire airways map and poster (incorporating 'Speedbird') and a mobile exhibition 'to tour the British empire in a railroad car'. F. MacCarthy, 'The fiery stimulator', *Guardian* (18 March 2006).
37 *The Times* (6 December 1935); *Imperial Airways Gazette* (June 1936); *Art and Industry*, 21 (September 1936), 110–13; le Corbusier, *Aircraft* (London, 1935).
38 *Straits Times* (Singapore) (28 April 1936; 21 September 1937).
39 London School of Economics (LSE), Brittain, 12; *Flight* (13 February 1936); (17 June 1937), p. 599; (22 July 1937), p. 103; *Sphere* (3 October 1936), p. 11; *Saturday Review* (18 September 1937), p. 187; Pemberton, 'Development', p. 87; *Imperial Airways Gazette* (October 1937, January 1938).
40 BAAM, AW/28/254: Memorandum to the *Imperial Airways* Board, 17 May 1938.
41 *Canberra Times* (28 December 1938), p. 4.
42 BAAM, AW/24/216: *Imperial Airways* Board Papers, 17 March 1936.
43 NA, INF 17/51; *Air* (December 1938), pp. 12–13.
44 *Flight* (18 August 1938), p. 147.
45 Hudson, *Air Travel*; Hudson and Pettifer, *Diamonds*.
46 *Air Review* (February 1936), p. 21; *Aeroplane* (14 September 1938).
47 *Sphere* (30 June 1927; 2 April 1932); *Empire Review* (March 1932); *Aeroplane* (4 May 1938).
48 *Aeroplane* (6 September, 22 November 1933); *Spectator* (17 November 1933).
49 *Popular Flying* (January 1936), p. 574.
50 BNIS, *British Paramount* (17 December 1931); *Sphere* (26 December 1931), p. 507; (24 December 1932); (5 January 1934), p. 7.
51 Previously, after a French railway crash delayed service between Paris and Brindisi, Powell made up thirty-five hours by 'flying night and day . . . in a thrilling dash between Cairo and Karachi'. *The Courier-Mail* (Brisbane) (2 January 1934), p. 14. *Imperial* would not have approved irresponsible flying.
52 *Bystander* (25 May 1932; 24 May 1933); reproduced in R. Prior, *Flying: the Golden Years* (London, 1991), pp. 96, 119.
53 H. Golding, *Wonderbook of Aircraft* (London, 1930).
54 *The Times* (31 July 1928), p. 9.
55 D. D. Nye, *American Technological Sublime* (Cambridge, MA, 1994).
56 *Spectator* (22 May 1926), p. 859.
57 *Sphere* (13 June 1931), p. 470.
58 Cluett, et al. (eds), *Croydon*, p. 2.

59 M. Grieff, *The Airport Book* (New York, 1979); J. Bamford, *Croissants at Croydon* (Sutton, 1986).
60 *The Times* (3 May 1928); Cluett, et al. (eds), *Croydon*, pp. 30, 32.
61 P. Sassoon, *The Third Route* (London, 1929), pp. 14, 18. *Aeroplane* (28 December 1938), p. 857.
62 Makin, *Swinging*, p. 169.
63 R. Coke, *Passenger by Air* (London, 1937); *Aeroplane* (4 May 1932), pp. 804, 806.
64 J. Cannon (ed.), *Oxford Companion to British History* (Oxford, 1997); Higham, *Britain's Imperial Air Routes*, p. 121; R. Finch, *The World's Airways* (London, 1938), pp. 22, 68, 82.
65 *The Times* (31 March 1926).
66 J. E. Morpurgo, *Barnes Wallis: a Biography* (London, 1972).
67 A. E. W. Salt, *Imperial Air Routes* (London, 1930), p. 24.
68 *Manchester Guardian* (3 July 1936), p. 10.
69 K. Munson, *A Pictorial History of BOAC and Imperial Airways* (Shepperton, 1970); Higham, *Britain's Imperial Air Routes*; *Aeroplane* (31 March 1937); *Vogue* (14 June 1939).

9

Imperial passages

Pilots, politicians and propagandists all publicised the possibility and desirability of imperial flying. In the 1930s the topic also received increasing coverage beyond narrow aeronautical circles. Independent newsreels, documentary films and slide shows were a prominent part of broad visual publicity about imperial flying.[1] There were references to the subject in a December 1937 lecture on the Sudan to the Royal Empire Society, for example. The speaker, a retired member of the Sudan Forestry Service (and a great-nephew of Gen. Gordon), showed lantern slides of *Imperial Airways* air routes, air mails being unloaded at Khartoum and an aerial view of the Croydon air complex. He showed the same slides at dozens of schools around Britain during a 1932/33 tour. If they spotted it, pupils may not have been aghast at the overlordship in the photograph of an Assistant District Commissioner and the wife of the local doctor supervising Sudanese men clearing the landing ground at Li Rangu (250 miles west of Juba) in 1933/34.[2]

British newspapers and magazines, several radio broadcasts, and vigorous book publishing for children and adults contributed accounts about imperial aviation and flying experiences. The larger daily and weekly newspapers produced a considerable amount of publicity. Several had appointed air correspondents by the 1930s. Among them were men who had acquired their interest and knowledge of aviation in the RAF. Others had experience as private pilots. William Courtenay calculated that he wrote 250 columns of air news in the London *Evening Standard* annually between 1932 and 1936 in addition to his other aviation activities.[3] *The Times* coverage included complete texts of the *Imperial* Chairman's annual reports, which, by the mid 1930s, took up two-thirds of a broadsheet page. Narratives of imperial flying were embedded in many such reports. Flying imperially, and reading and writing about it, framed and defined the experience.

Serials

Imperial Airways never urged anyone to 'write imperially'. No serial publications published in Britain in the 1920s and 1930s were devoted exclusively to Empire aviation, but many magazines and journals about aviation and flying carried articles about Empire air routes, services and flights. The encyclopaedic *Air Annual of the British Empire* represented specialised serial publishing about Empire aviation most voluminously. Founded in 1929 and edited by Sqd. Ldr C. Burge, the first of ten annual editions was a weighty 700-page tome. It may not have been a bestseller but it would have helped give Empire aviation a profile, even if only in the industry and in well-endowed and progressive public libraries.

The first edition of the *Air Annual of the British Empire* opened with a foreword by Sir Samuel Hoare, then Secretary of State for Air. There followed several chapters in a prominent section entitled 'Empire Aviation'. Leo Amery (Under-Secretary of State for Colonies) penned a piece on 'The Empire and Air Communication'. Sir Harry Brittain wrote about 'Empire Air Policy' and de V. Robertson examined flying in the Dominions. There were two chapters on air survey and Empire development. In a chapter entitled 'Aviation and the Empire', Sir Charles Wakefield recorded the benefits of increased air travel: diminishing the likelihood of war (even outlawing it), strengthening Empire and sensitising residents to its uniqueness. Subsequent editions of the *Annual* contained less extensive but no less enthusiastic commentary on Empire aviation. In the 1933/34 edition, for instance, Woods Humphery noted that Britain's Empire airline had the task to discharge an age-old duty and privilege: 'the air is to be Britain's heritage no less than the sea'.[4]

Monthly and weekly specialist aviation periodicals also publicised aviation and 'airmindedness'. Among these was the *Journal of the Royal Aeronautical Society* that catered mostly for aeronautical engineers. *Flight* (the official weekly organ of the Royal Aero Club) carried material of wider general interest. Periodicals devoted to enthusiasts rather than scientists and constructors included *Popular Flying*. Aimed at an adult market, the 6d monthly reached a monthly circulation of 24,000 within two years of its launch in 1932, and 32,000 in 1938. W. E. Johns, a wartime pilot and then aviation illustrator and story writer, edited the magazine until 1939, adding the sister weekly paper *Flying* in 1938. His aim was to counter the flood of American 'aviation pulp' and build an airminded culture: Britain must fly, he wrote, 'because the nation which does not fly will be left behind'.[5]

Johns became (in)famous for creating 'Biggles' (Maj. James

Biggleswort h, DSO), a redoubtable young pilot in boys' stories. The caricature chipper, upper-class Englishman appeared in several London magazines (including *Modern Boy*, for which Alan Cobham also wrote). In time, Johns was criticised for racism and sexism, and the story books were swept off some library shelves: some readers had tired of the myths and stereotypes used to legitimate and sustain the Empire.[6] Johns' defenders have argued that his (later) flying stories contain explicit reminders that Africans and others 'are no more savage than their Western counterparts' and that his stories often challenged young readers 'to question the benefits of Empire'. Late in the day, Biggles did indeed recognise the rights of indigenous people, and did regret the decimation brought by the habits and diseases of colonisers.[7] No critics remarked on the imperial values and projections surrounding the act of flying itself.

Modern Boy, like Saturday magazines such as *The Ranger*, included serialised thrills-'n-spills flying tales, mostly about fighting the Germans, Arabs, villains, bandits or the elements. Excepting for record-breaking flights, civil flying was too anodyne to sustain regular magazine sales. Among occasional accounts of commercial flights in the second half of the 1930s, two cover features in *Flying* narrated Empire air journeys by *Imperial* flying boat: in addition to the author, the African flight carried the wife of a Tanganyikan farmer, a widow on her way to Tanganyika to get remarried, an *Imperial Airways* ground engineer based at Lagos and a Swedish airline employee.[8] An edition of *Popular Flying* carried a lengthy illustrated account by Lily Jameson of her 8,323 mile *Imperial* flight home from Singapore to London after spending two years in Malaya. Her otherwise standard rendition noted sightseeing in Delhi while the aircraft waited for delayed incoming mails, and the radio playing comforting 'music from England' in the stillness at Sharjah fort.[9]

In the same paper, Oliver Warner wrote up his *Imperial* flight from the Sudan to London in 1936. The Greek man in charge of petrol stores at Kosti told him that he was the first-ever passenger to depart from there. A broken fuel nozzle delayed take-off in 104°F heat. Flying in the company of an eighty-one-year-old lady returning from a pleasure trip to the Cape, Warner felt elated by a trip 'of positively staggering interest'. The cabin was not too noisy to prevent conversation. Nobody became airsick; no paper bags were issued. His flight had Warner reflecting on the different flying cultures in England and in the Empire. In the Sudan, he wrote, most people knew a lot about aeroplanes. Some could fly themselves and, if they could afford it, nearly all proposed to return to England by air for their leave. Conversely, citing Aldous Huxley, he noted that aeroplanes still meant little to 'the ordinary

Englishman' who lived and moved as though the Wright brothers and Bleriot had never existed.[10]

Another account in *Popular Flying* was by Mr G. M. Fleming, a pilot for Wilson Airways and Goldfields Air Mail Service. One of his anecdotes was about his reception after a forced landing in East Africa: the 'whole tribe' turned out, he said, 'minus clothes', and the chief asking 'why the big bird carried two bwanas in its stomach'. Dress – or nakedness – was once again invoked as a visible register of difference and marker of 'civilisation'. As ever, the term 'tribe' was used loosely to refer to any collective of 'primitive' people without a known name.

There was even more excitement in connection with Fleming's attempt to fly himself and an American friend from Nairobi to England. An oil leak forced them to land; the nearest help was at a cotton ginnery four hours' walk away. They sent an African there with a note requesting help and oil. In the interim they surveyed the ground and argued about the price of engaging twenty Africans to cut the grass and clear the surface. A start was then made on preparing a 20 x 300 yard runway: the Africans uprooted bushes, cut elephant grass and filled in pig-holes. An English-speaking African clerk arrived with ten men armed with spears; all European men were away on safari. He had a bowl of rice and a mosquito net delivered, and sent a message to Juba, a three-day walk. The second day was spent removing an eight-foot-high anthill. An African was kept busy boiling Nile water in an oil can so that it could be drunk. A camp fire kept animals away. Oil arrived at the end of the third day. 'Practically the whole tribe' arrived to see off the two visitors (Fleming thought it important to note that only the chiefs wore clothes). The hiatus was the first of many; after eleven forced landings the two abandoned their plane in Italy.[11]

In 1934, a magazine of commercial air travel was launched, also under C. G. Burge's editorship. His editorial in the first issue was about Britain as chief partner in a worldwide brotherhood of nations. Aviation, he wrote, would consolidate the British Empire, revive Empire trade, improve international understanding and spread peace in the world. The subtext was the exercise of imperial power. Disraeli had enriched Britain and the Empire by buying shares in the Suez Canal; Burge called for 'men with similar vision who will exploit the air'. Contributions to the magazine included articles on aviation and big game in Africa. A survey of epic flights was the opportunity to romanticise the air and defend publicity against attacks of vulgarity. Burge argued that publicity had an honourable history, driving achievement even in classical times, and that prestige marched with publicity.[12]

The *Aeroplane* appeared weekly under the stewardship of the eccentric and irascible C. G. Grey throughout the period 1919–39. Covering military and civil aviation, its circulation was small (approximately 7,000 in 1929 and 10,000 in 1934), but what it lacked in numbers it made up for in influence: Fleet Street scanned each issue as soon as it was printed. At least one youngster spent more on his bus fare fetching a copy on the day of publication than he did on the sixpenny cover price. The paper boosted his self-esteem, he recalled.[13]

Letters about Empire aviation appeared quite often in the pages of the *Aeroplane*, as did anecdotal accounts of Empire flights. The magazine advertised lectures relevant to imperial aviation, and several were summarised. Readers were told about the two talks on Empire air routes given by *Imperial*'s Publicity Director, Mr F. Snowden-Gamble, to the Institute of Civil Engineers and to the Royal Aeronautical Society. Similarly, the magazine publicised the second Brancker Memorial Lecture given by Handover in 1936 on 'Some Aspects of the Organisation of Empire Air Services'. It also printed a summary of the fourth Brancker Memorial Lecture given by the Assistant-Secretary of the Post Office on 'Twenty Years of British Air Mails'.[14] Empire flying was a popular lecture topic even beyond London. Following fifteen other lecturers in a prestigious annual series devoted to topics of contemporary imperial interest such as banking, food supply, tropical development and migration, Lord Londonderry, the ex-Secretary of State for Air, spoke about Empire airways in a 1936 lecture at University College, Nottingham. He would have been able to draw on his own experience of using *Imperial Airways* to reach RAF bases in Egypt, the Middle East and India during the Christmas 1933 parliamentary recess. His seven-week trip covered 16,000 miles.[15]

For a time, the Air League of the British Empire published a monthly magazine titled *Air and Airways*. It had a narrower civil aviation remit than *Aeroplane*, but also printed news items and full or abbreviated versions of parliamentary debates. In the years 1931–33 the magazine carried ten or so short pieces in which authoritative writers tackled aspects of Empire aviation. The magazine's predecessor, *Airways*, started publication in 1924 and was marketed as 'The Only Air Travel Magazine'. On the cusp of Empire aviation, pieces of imperial interest appeared among its editorial, articles, news items and trade advertisements relating to commercial and private aviation. It carried feature supplements on the Australian flights of Cobham and Hinkler, articles on air services to India, the airway across British Africa, and reports on the long-distance flights of Cobham, Bailey, Johnson and the *R100*. Colonial images appeared in an article on the Northern Rhodesia air survey, where photographs of two Africans (called 'darkies') were shown developing film in a darkroom at N'Changa, and more than twenty Africans were pictured demolishing an ant heap at an emergency landing ground. An ex-member of the 'East African Intelligence and Native Administrative Service of Tanganyika' contributed photographs of African women clearing a landing ground and African men building an aircraft hangar of poles, mud and thatch. The *Imperial Airways* Africa timetables often carried an imperial theme. The cover of one paid for by the manufacturer whose engines had taken Lady Bailey across Africa showed an African warrior gazing at an engine block placed in the middle of a stylised African landscape over which a light aeroplane flew in cloudless sky.[16]

The textual commentary on Empire aviation that appeared in newspapers and magazines was informative and often persuasive. Arguably, however, visual images beat the printed word as the most forceful representation of Empire aviation progress. The world air route maps published periodically in *The Times* and in *Aeroplane* and *Flight*, as well as in aeronautical books, transmitted a powerful message about tenancy of the sky and about imperial communications. The map of existing and planned air routes in the Empire Press supplement of *The Times* on 31 May 1930 gave a bold impression of an Empire linked together by means other than those deployed by British shipping and cable companies. These had long boasted about their global reach in maps printed in exhibition and trade supplements in the newspaper. The cartographic device was no different from that used to depict any transport route, but air route maps comprising chunky and darkly inked lines erroneously suggested a continuous physical presence like a canal, road or railway. In reality the Empire air net was only gossamer. No more than three or four *Imperial* aircraft were airborne at any one time. The tiny, foreign objects barely ruffled the air, and cast only small and fleeting shadows.

Had matters of Empire aviation been confined to the pages of aviation periodicals and newspapers, 'airmindedness' would have penetrated less deeply than it did. London societies catering for people whose interests were regional (African or Asian) rather than aeronautical helped to spread awareness. Speakers at these and other society meetings would often be invited to publish their talks in the house journal. One of many examples is the lecture given to the dominion section of the Royal Society of Arts in December 1928 by Col H. Crosthwait, once Superintendent of the Survey of India. In the Society's *Journal* he could publicise more widely his opinion that the slogan 'buy imperially' was undermined by the failure of the Empire to meet consumption requirements in the United Kingdom. Despite the variety of climate and soil in the Empire, only about thirty-eight per cent of food and tobacco imports into the United Kingdom derived from Empire resources. In a sentence striking for its possessive pronouns and final verb, Crosthwait identified 'considerable room for the development of our resources before the Empire can nearly supply all we want to take'.

In addition to arguing that aerial survey could help turn dormant raw materials into resources which could make the Empire more self-sufficient, Crosthwait linked airborne survey to unemployment and work relief in Britain: it might assist population redistribution, declining emigration and Empire settlement. In his lecture (which prompted a glowing editorial about 'Air Survey and Empire Development' in *Nature*, the scientific magazine), Crosthwait suggested that bringing large tracts of idle land into production offered the prospect of sustainable work away from overcrowded Britain. Aerial surveys and maps would facilitate economic development by allowing accurate assessment of the Empire's forests, minerals and agricultural resources, and would enable land to be allotted to [White] people who were anxious to occupy it. Surveys would also help in the alignment of roads and railways, and in the construction of irrigation and hydroelectric schemes. Later, speaking to the Royal African Society about air surveys in East and Central Africa, and then publishing in its *Journal*, Crosthwait added that aerial survey could play an important part in the disposal of public lands, and therefore assist public welfare and social progress.[17] His presentation contained no notion of debilitating land alienation or dispossession.

Articles about the role of flying in colonial surveying, reconnaissance, mineral prospecting and administration appeared in the non-technical periodicals *World Today* and *Crown Colonist*. The writer of an article in *United Empire*, the journal of the Royal Empire Society, described aerial survey as 'a new gift of science to the Empire'. His florid argument for pressing aircraft into imperial service was that England could no longer afford to let the wealth of its 'torrid zone estates' lie dormant. Not only were aircraft a wonderful auxiliary for bridging gulfs which divided the members of the British family, but aeronautics was on hand to help 'idle hands' find work exploring hidden treasures and extract new wealth from soil and mine. A small and indistinct photograph accompanying the text showed Africans smashing a massive anthill to clear space for an aerodrome for the Northern Rhodesia aerial survey. Was it the idleness of Africans, expatriates or Britain's unemployed that the writer anticipated being harnessed for science and the realisation of buried treasure? Setting idle or meddlesome statesmen and politicians to work was another matter. Subscribers to *United Empire* were told explicitly that the Empire would be stronger and more united if politicians had less to do with it.[18]

Aerial survey was a favourite subject among organisers of lecture programmes. The blend of photography, aeronautics and economic progress overseas was a winner. Early in 1930 Lord Thomson spoke on air survey at a lunch meeting of the British Empire League and at another meeting in Cardiff. Delegates to the annual conference of the Geographical Association at the London School of Economics in the same year heard first hand from someone whose ten years' experience included surveying Venezuela's Orinoco Delta and Burmese forests and creeks. He had also flown surveys for copper mining, railway alignment, town planning and state boundary delimitation in Northern Rhodesia. Despite this impressive record, Maj. Cochran-Patrick told his London audience that only one-fifth of the Empire's fourteen million square miles had been surveyed on a regular basis. Science was waiting to complete the picture. Aircraft could do the job between four and five times as fast as ground survey, at a cost up to one-third less. Moreover, aerial survey minimised exposure of European surveyors to ground work in hazardous and unhealthy environments.[19]

Harsh physical environments were settings in which flying had obvious application for search and rescue missions. The case held good for European expatriates, but could also be used to highlight other benefits of British stewardship and the aviation it promoted. This public service facet of colonial flying was seldom publicised in Britain. An exception is the instance in which vaccine was flown from Nairobi into rural Kenya where an outbreak of smallpox had killed 470 people. Many more lives were spared than would have been the case had medicine been transported by road: the flight took two hours, the road journey ten days.[20] As its promoters knew well, the lesson was that flying was particularly valuable in places where other forms of transport were poorly developed. Aviation had an almost blank canvas in the African veld, the Australian outback and the Canadian shield.

Short articles about various aspects of Empire aviation appeared occasionally in several contemporary British magazines. Rather than draw the attention of a wide public to the promise and actuality of Empire aviation, the articles would have been reminders to an educated and prosperous readership. As an historical source the material reads repetitiously. Material appeared in magazines such as the *Illustrated London News* and in general-interest magazines such as *Blackwood's Magazine*, *Chambers's Journal*, *Discovery*, the *Empire Review*, the *Economist* and *Outlook*. *The Graphic* published a special supplement in November 1931 to mark the 'forthcoming completion' of the London–Cape Town air route. Magazines with a regional emphasis, such as the *Asiatic Review*, *Near East and India* and *Great Britain and the East*, published brief articles on Empire aviation. Papers also appeared in the journals of the Royal Central Asian Society, the Royal Institute of International Affairs, the Royal Society of Arts and the

Royal United Services Institute. Whereas in the 1920s the Air League *Bulletin* had been a lone messenger, by the 1930s Empire aviation had a broader press.

At some remove from aeronautical and Empire periodicals, the in-house *Post Office Magazine* gave some coverage to Empire aviation. In the mid 1930s it circulated to 170,000 employees. The cover photographs of several editions featured air mail handling and ceremonies at Croydon. Empire aviation was also (rather surprisingly) given some coverage in the *Field*, a horsy, Home Counties weekly. In a new departure in the 1930s, the magazine started a regular section entitled 'Travel by Land, Sea and Air'. Sir Donald Campbell (the 1935 world land-speed record holder) was appointed the motoring editor and Lord Sempill the aviation editor.

Sempill's credentials were several: he was well informed, could communicate with non-specialists and was inspiring. He was the immediate Past-President of the Royal Aeronautical Society (1927–30), had published a small book entitled *The Air and the Plain Man* (in the 'Library of New Ideas' series) and was a sought-after public speaker. His stock-in-trade speech about 'airmindedness' was well practised by the time he spoke to boys and their parents at the City of London School prize giving in 1933. The senior pupils may have prepared for examinations using a new Economic Geography textbook that mentioned air transport in a chapter on Empire communications. It suggested two practical exercises: plotting existing and planned Empire air routes, and describing an aeroplane route from Croydon to Delhi.[21] A map would have gone some way to confirming what the Professor of Economic and Regional Geography at London University argued in his newly published *Political Geography of the British Empire*. The Empire, he wrote, had a greater interest than any other 'state' in the development of 'reliable long-range air-vessels', for no other power had its territories separated by such vast distances.[22]

Mindful of his weekend adult readership, Sempill wrote extensively on light aircraft, private flying and air touring, air clubs, gliding, aerodromes, air navigation, the minimum age of trainees and pilots, and the use of aircraft for domestic policing (as 'aerial bloodhounds') in Britain. Yet, as was the case in the unsigned 'Air Progress' section in the weekly 'Round the Empire' survey in the *Saturday Review*, his articles reflected the way in which domestic and overseas flying intersected, offering each other mutual support and advancing the aviation capability of the British Empire. For example, a piece about the aeroplane and colonial development noted that throughout the Empire light aircraft were breaking down the isolation of settlers' homes, taking some of 'the amenities of civilisation' to their firesides, and playing an important part in promoting a feeling of unity among widely scattered pioneers. Sempill did not neglect to alert readers of the *Field* to the way aeroplanes allowed game to be tracked easily and studied at close quarters. On another occasion, he criticised the way in which air pageants and air shows glorified speed, spectacles, stunts and militarist tendencies. Instead of underlining the safety, reliability and peaceful uses of aviation, these events 'pandered to the modern craving for sensational thrills'.[23] Conservative country readers would have grasped Sempill's disquiet.

Sempill's magazine column tackled Empire aviation themes head on as well as indirectly. As in his book, so too in his column he lauded the British aircraft industry, 'an imperial asset of the greatest magnitude'. He tackled the meaning of Empire Air Day, insisting that the prosperity and safety of the British Empire were indissolubly linked with aeronautical progress. Pursuing the theme, he rehearsed the convention that aviation propaganda should bind aviation as closely to national life as maritime supremacy did in the past. He wanted flight to be part of British culture: 'the aim must be to make air-sense as much part of the British character as sea-sense has been for so many generations'.[24] Sempill also kept a keen sense of the link between national prestige and success in aviation. He warmed to the Everest flight, but cautioned that the Melbourne air race should not be allowed to degenerate into a happy hunting ground for publicity-seeking sensation mongers. In his view that trend had to be balanced by the consideration that the prestige of British aeronautical design and construction had been dented by losing the air speed and endurance records, and by the delay in linking Britain by air to Australia.[25]

In the 1930s boys' annuals began to carry stories with an aviation theme. The *Modern Boy's Book of Aircraft* competed with the *Collins Aircraft Annual* which, in 1931, featured contributions from Cobham and Lindbergh. At least six flying stories for boys were published for the Christmas gift market in 1933. The *Times Literary Supplement* book reviewer sensed a standard recipe, but anticipated an unchanging appetite and unimpaired digestion. The recipe's basic ingredients were youths of extraordinary athleticism and skill, highly powered aircraft and 'villains of deepest dye'. The best stories included a chase of alarming intricacy and a spice of mystery. Villains had to be dispatched, and the whole text needed serving piping hot with bloodthirsty sauce.[26]

A formulaic heroic air rescue featured in the 1933 edition of the *Empire Annual for Boys*. One of many adventure stories in the issue told of an army major who, accompanied by his nephew, flew his light aircraft from London to Cape Town to recover some stolen jade figures. Discovering that the (Cockney) thief had flown his loot deep into Africa, they headed north with an African guide. With requisite suspense and struggle they retrieved the idols. An aeroplane arrived in the nick of time to rescue the three Europeans from a 'wild tribe' who watched in awe as it took off and faded from sight. The African guide and other native followers were left to walk back to Cape Town in a journey that took 'many moons'.[27]

The 1934–35 edition of the *Boy's Own Annual* (edited by a Fellow of the Royal Empire Society) carried a story about an experienced pilot and a trainee who, on a mail-only flight from London to Bombay, made a forced landing and warded off a gun attack by between thirty and forty Arabs. The story-line was given its urgency by writing which announced that in distant, sunny Bombay, 'countless British exiles in civil and military service waited for news from home, news that would give them courage and endeavour, that would spur them to carry on with the task of Empire building, knowing they were not forgotten by friends at home'. The young pilot's heroics were about his determination that the mail must get through at all costs; he 'would not be the first to give up his life for an ideal'. 'Over and over in his mind in a frenzied hysterical procession went the words, London to Bombay, London to Bombay.'[28]

Weaving an Empire theme directly into aviation stories was not difficult: unknown and imagined wilderness overseas was a setting ripe for adventure. The author of a story serialised in 1933 seized the opportunity. Running against the current of adventure and aviation stories for boys, the plot involved girls and appeared in the weekly *Schoolgirl* magazine. This feminised 'dollop of diluted Rider Haggard' was a story about the flying sisters surnamed Fortune. The episodic and fractured tale worked with a range of African caricatures. The sisters, orphaned with their brother in an unnamed part of Africa, have to tackle villains wanting the plans of a gold mine. In the 'savage' continent the girls confront wounded lionesses, hostile giant apes, mischievous and highly organised bands of chimpanzees, and two warring pygmy 'tribes'.

The story includes a line drawing of fourteen-year-old Joan, who stands in African grassland at the head of a line of 'pygmies' carrying shields and spears. The Africans wear grass skirts, anklets, ear rings and head bands. Joan is fully clothed in flying kit. She is taller by far. She is white-skinned and her name is English but, lest there be any doubt, her bi-plane (*Sky Queen*) is a distinctively English design. A small and alert terrier adds a domestic English touch. One of the 'pygmies' presents her with a shield and spear. Joan does not offer her dog, flying cap or goggles in exchange. This is not trade; it is tribute. This is what infantilised Africans do when White people appear. Joan's only reciprocal gesture is an adult and distancing pat on the head for the gift-giver. To exclaim, she at once became 'the proudest pygmy in the tribe!'[29]

Books

All publicity about British aviation was good publicity as far as aviation authorities were concerned. Popular books eased their own burden somewhat. Brancker made this plain in the foreword he wrote to Harper's 1930 book *The Romance of a Modern Airway*.[30] It says something about the way aviation (and romance) was understood that Harper's newest book was part of a London publisher's 'Romance Series' which included titles on the heavens, submarines, archaeology, navigation, ocean liners, railways and the civil service. The new gods were technology, discovery, speed and management. They had to be appeased by contriving a nostalgic romanticism. Harper's book depicted the raw material of flight (photographs of aircraft interior cabins and landing grounds) and captured its modernity by emphasising contrast. A photograph of an *Imperial* landplane was shown behind camels standing in the foreground. A seaplane was shown taxiing behind a sailing ship. The international dimension of flight was frozen in a monochrome photograph that showed the smiling Sultan of Zanzibar alighting from an aeroplane after a flight over London. On another page, a group of 'vastly impressed' 'Red Indians' was pictured posing at Croydon. A picture of a young boy standing in front of a scale model of an *Imperial* airliner indicated that a new phase of toy-based acculturation was starting.

Harper had no monopoly on popular aviation publishing. Maj. Salt's 1930 book *Imperial Air Routes* was judged by a reviewer to be among the most important post-war books on aviation (and not just because of the preface that Brancker must have jotted down in a matter of minutes). It was by now common dogma that to hold the Empire together transport construction and operation had to reach and maintain the same position in aviation as in seafaring.[31] Cobham made three contributions to the book-fest with personal accounts of his Australian and African flights. He dictated one at breakneck speed over a weekend.[32] Other pilots published their African flying experiences too.[33]

Empire air adventure and 'aerial conquest' were enjoying a renaissance. The 1932 publication, *Down Africa's Skyways*, was a retrospective account of several pioneering flights to the Cape.[34] Although the writing was criticised for having 'too much of the headline, the startling adjective, the whirr of the cinematic camera, the click of the newspaper telephone', the reviewer conceded that the book's harmless exuberance was as much a part of the contemporary scene as aviation itself. Sir Harry Brittain evidently did not regard the text as the 'modern Hakluyt' that the African airway warranted.[35] Harper and Brenard's *The Romance of the Flying Mail* (1933) was worse, if C. G. Grey is to be believed. Grey condemned the survey as unimaginative, incomplete and encumbered with 'abominable platitudinous plantigrade journalese'.[36]

In the early 1930s the market for books about British aviation appeared inexhaustible. An author who had the handiest of surnames (and had just written two popular accounts of British broadcasting) dashed off a text. Sociable and prolific, Sir Harry Brittain's aeronautical interests and contacts had grown when he chaired the civil aviation section of the London Chamber of Commerce, and when he was in the air transport section of Britain's delegation to the 1931 Washington conference of international chambers of commerce. Sir Harry researched his book through the expedient of *Imperial Airways* giving him and his secretary-friend and writer, William Makin, seats on an aircraft delivery flight to Africa in early 1933. A long delay at Atbara for mechanical reasons stretched the outbound flight to three weeks (and may have explained the passenger waiting lists at every airport). The trip was considerably longer than Brittain's experience of inaugural flights (twenty-two hours) on the *R38* and *R100* airships. The passage out became an extended essay on colonial readings of place and people, highbrow networking and extravagant entertainment.[37] In Cape Town, Brittain was persuaded to give a forty-five-minute talk to both Houses of Parliament on 'The Romance of Aviation'.

Sir Harry's 1933 book *By Air* had been published only ten days when it went into a third edition. He recalled receiving kind and enthusiastic newspaper reviews and letters of encouragement from air enthusiasts throughout the Empire. A reviewer in the *Times Literary Supplement* regarded Brittain's analogy between *Imperial Airways* and the British East India Company 'rather loosely enthusiastic', but felt that the factual record made a fine tale and that propaganda was still needed. Brittain's publishers asked him for another book, one that would inspire the youth of the Empire. His answer was *Wings of Speed*, published in 1934. Both books opened with ringing endorsements of *Imperial Airways* by Lord Londonderry, the Secretary of State for Air: it was, he wrote, a great organisation of which the Empire should be proud.[38]

A government minister could be expected to make glowing remarks, or at least to paper over cracks. What counted was not just the truth, but the impression which was created. Brittain's youthful readers would take as fact the opinion of an important-sounding person. References to greatness and pride would feed their fantasies no less than the qualities of individual heroism and comradeship that the Secretary of State discerned at the heart of imperial aviation. Boys would have been more impressed by these qualities than by the 'patient enterprise and sound craftsmanship' that they were told the Empire had previously depended on for mastery of the sea, and on which it still relied for its future destiny in the air. Boys were also unlikely to read or be put off by one hostile reviewer's opinion that *By Air* – and presumably *Wings of Speed* even more so – slid too easily between narrative, gossip and statistics, and that there were just too many birdmen blazing the route in droning or roaring air argosies. A more sympathetic reviewer wrote that what the book lacked in poetry it made up for with 'dollops of happy enthusiasm and immense gusto'. It drew on an unusual array of facts and figures. Overall, it was 'a very healthy work of propaganda'.[39]

Brittain compressed a great deal into his writing. In just a few casual remarks across a handful of pages he made sweeping statements about the favourable social impact of Empire aviation, for example. At Khartoum, where once Gen. Gordon waited helplessly for news, British people were now in rapid communication with London. English exiles gazing across the Nile could satisfy their cravings for the chit-chat of London life. Nairobi had almost become a suburb of London. The Uganda Game Reserve was as popular with zoologists and zoo-gossip writers as England's Whipsnade Zoo. Soldiers flying back to the Empire from home leave arrived in better shape than if they sailed. Brittain also worked with themes of order and progress. Over Africa, the engine roaring in the sky tore across a map of carefully designed political boundaries. Were there to be any exciting incidents at African outposts which might have unseated a British government in the past, they could be attended to speedily after a brief flight.[40]

Wings of Speed was not the first or the only writing for a youthful readership. In 1931 a non-technical aviation history for boys appeared under the title *The Book of the Air*. In 1933 the publishers of the *Boy's Own Paper* released a collection of diluted accounts of pioneer flights under the title *Twenty-Six Flying Stories*.[41] In similar vein, a 'Great Adventure' series of books included one called *A Million Miles in the Air*. A reviewer judged it 'a noteworthy piece of propaganda for British aviation and its pilots'.[42] The next year a chapter about the 1934 air race to Australia appeared in a volume entitled *Epic Tales of Modern Adventure*.[43] W. E. Johns published a collection of twenty accounts of 'thrilling flights' from *Popular Flying*.[44] In the mid 1930s the Nelsonian Library for 'boys and girls of all ages and tastes' published three aviation titles. Narratives of pioneering flights mixed with accounts of organised civil aviation. One text urged readers to picture *Imperial* as the 'long-distance air highway of the British Empire'.[45]

Such was the demand for information about flying that by 1934 the *Wonder Book of Aircraft* was in its eighth edition. Brenard's brief essay on 'Spanning the Empire By Air' lurked between the eight colour plates and 250 illustrations in yet another volume edited by a Fellow of the Royal Geographical Society. The piece told of a rest house fort 'in the heart of desert country infested by hostile Arab tribes'. Along the African airway ('undoubtedly the most romantic of any air-line in the world'), Brenard wrote that airway staff waged 'constant' war with climate, plants and wild life.[46]

Imaginative aviation writing was also booming. In the mid 1930s the number of flying adventure stories warranted special coverage in the *Times Literary Supplement*.[47] Storybooks that had aviation as a theme, and pilots for heroes, proliferated. Flying stories became 'a minor literary genre'. One publisher ran a special series entitled 'Air Adventure'. Another produced over seventy stories in a series named 'Ace'. Typically, stories idolised the exploits of fighter pilots. A few were quasi-factual. Their style and content were unremarkable. Their informational and literary value was surpassed by their propagandist merit.[48]

The writing struck an imperial note even if it did not tackle the subject explicitly. Pollard's books for boys are a good example. The collection had no obvious imperial resonance but did contain themes of heroism, service, gallantry, courage, exploration and pioneering. Many boys would have been seized by Pollard's citation of the remark Churchill made in 1919 when acclaiming Alcock and Brown's trans-Atlantic escapade. Flying, Churchill suggested, projected the best of a glorious national tradition into the present: 'we have preserved as a race the audacity, the courage, the physical qualities of the old heroic bygone times . . . our civilisation has combined the science of the twentieth century with the virility and love of adventures of the knights of old.[49]

Two well-known writers of boys' adventure stories, John Westerman and W. E. Johns, contributed other books which worked with these themes and presented Empire flying as honourable. The *Air Record Breakers* (1937) is about a search for precious stones in the South Pacific. Leaving England, two 'chums' fly themselves to Cape Town, Brazil, New Zealand, India and back to England under the pretext of testing a new seaplane and fuel for the Air Ministry. On the way they foil an attempted hijack between Bombay and Basra, and complete a record Sydney–England flight in four days.[50]

The map and meaning of Empire airways were etched similarly by Johns. His young readers were 'always being reminded that planes link Britain to the furthest corners of the Empire' and about conspiracies to destroy the Empire.[51] In *Biggles in Africa* (1936), three boys uncover Turkish drug farming in (fictional) 'Insular' between Malakal and Juba while searching for a wealthy Englishman's son who disappeared on a solo flight to Cape Town. The characters include a White hunter, a one-eyed 'half-breed' manager of a disused *Imperial Airways* emergency landing ground, African 'savages' and a grotesque 'witch doctor'. The tale ends with dinner back in orderly, safe, righteous 'civilisation' at the Savoy Hotel for the heroes of the rescue.[52]

Adults were also served well by sprightly aviation book publishing throughout the 1930s. It may only have been a mild exaggeration to claim in 1936 that flying had generated more writing in proportion to its age and size than any subject except murder and love.[53] The five aviation books published before Christmas 1935 for adult readers (including two contributions from Capt. N. MacMillan, the President of the National League of Airmen) tapped and fuelled a growing public 'airmindedness'. The deluge of aviation publishing continued up to the threshold of the Second World War when, for one, Amy Johnson was published for an adoring public.[54]

Dower's 1934 account of his intentionally slow flight in a light plane from London to Madras two years previously tried to avoid imperial moralising, but let slip remarks about the vileness of Mesopotamia and the evil reputation of Persian Gulf sharks, bandits and government officers.[55] Charles Lorne's fictional *Air Liner* (1934) barely raised the tone. Set on a flight from Cape Town to Croydon via Cairo and Brindisi, the 'whodunit' featured an odd cast of characters, albeit their skin pigment was a 'safe' white. Reviewing the book in the *Spectator*, the distinguished South African writer William Plomer read the cast of passengers as 'a vamp, a neurasthenic eloping with another man's wife, a restless Jew, a retired hospital matron, and a blatant vulgarian and scoundrel'. In reality, snooty passengers who drew a fine line between eccentric and motley would have been dismayed at the idea of having to sit among such people on the Empire services. Travelling in a confined space for several days among mysterious people may not have been utterly dissimilar to sitting next to strangers on a London omnibus, as another reviewer remarked, but it was a little unexpected.[56]

Systematic aviation studies that had Empire interest included Monk and Winter's two contemporary histories, and Finch's 1938 survey of the world's airways.[57] Reviewing Lord Thomson's 'Air Facts and Problems', Francis Yeats-Brown added his own thoughts about predestiny: 'we have the men, machines and money to keep our appointed place in the air as we did by water'. He ended by citing two lines from Tennyson's poem 'Hands all Round': 'Pray God our greatness may not fail / Through craven fears of being great'. Knowledgeable readers would have recalled with pleasure the poem's imperial sentiments about an England under Indian skies and about keeping the glorious English Empire whole. They might have amused themselves by substituting aviation terminology for Tennyson's maritime vocabulary: 'We sailed [flew] wherever ship [plane] could sail [fly] / We founded many a great state'.[58]

The subject of aviation was tilled industriously in interwar publishing, but its quality was uneven and its veracity was compromised in the interest of a powerful message or a good yarn. It was a long while before there was any literature that bore comparison with Joseph Conrad's and Herman Melville's sea writing. And how galling that no writer in English could match the Frenchman Antoine de Saint-Exupéry,[59] until the American, Anne M. Lindbergh, produced her 1938 narrative *Listen! The Wind*. Agatha Christie's mystery *Murder in the Air* (1937) was stirring but it reached only modest literary heights. The art had still to be learnt of how to capture the romance of the air without straining incredulity. It was all too easy to snap at the 'slipshod literature' accompanying the development of aviation. One review of a clutch of aviation books released for pre-Christmas sales in 1936 noted that although publishing continued 'with ceaseless gusto' it was hard to imagine who read the books. The point was well made that experiencing flight just once did not automatically provide credentials to write about it. Irrespective of how sensational it was being airborne, and as riveting as the spectacle of flight was, commentators found that writing about aviation was rarely other than prosaic and clichéd. Poetry and flight met only occasionally on the printed page in a quiet panegyric.[60]

Most writing about flight was strident and was about excitement. Some focused on policy and numerical observation. Some comprised sequential descriptions (and photographs) of sights from the air, and of refuelling and overnight stops. Coke's 1937 book about, among others, his trips on *Imperial*'s inaugural passenger services from Cape Town to Cairo (1932) and London to Sydney (1935) exemplifies the genre.[61] There was almost no literature that made flight part of any aesthetic. Several critics commented on the failing. In 1930, Francis Yeats-Brown lamented the absence of Britain's poet of the air. 'The airmen and airwomen of tomorrow', he said, 'must have a retinue of bards to celebrate their exploits.'[62]

For years there was nothing written which would prevent C. G. Grey from pronouncing most poetry of aviation revoltingly sentimental or unprintable. Whether or not aeroplanes deserved their own poetic was questionable. One writer thought that an artist in words was needed to capture the 'roaring surge towards the sky, the thunder of exhausts, the rhythmic drone of super-powered engines, the high-pitched whine of the radio at work, and the swishing silence of the plane's descent', or indeed, to go beyond these elements.[63] Another writer, disillusioned by flying, thought that modern aircraft and commercial flying did not lend themselves to poetry.[64] But poetry could certainly have captured the imperial imaginaries and sensibilities of flying. In 1943, an American author expressed astonishment that self-deprecating British aviation pioneers had reduced their efforts to workaday tasks. He found it strange that unlike the English literature celebrating the merchant marine, there was no soul-stirring writing about how the Empire's lifelines were lifted into the sky.[65] The hot air of rhetoric and propaganda was indeed finite.[66]

Radio

Unseen speakers added a new dimension to information and attitudes about flying imperially. While inter-war household radio ownership in England was increasing annually toward a figure of seventy-one per cent (8.9 million) in 1939, the subject moved gradually from news to feature item. Two new technologies grew mutually, both airing the Empire.

When he was Secretary of State for Air, Sir Samuel Hoare gave one of the earliest radio broadcasts about aviation. It was in November 1925. He began by noting how the aeroplane had revolutionised Britain's geographical position. Mentioning Blériot's unwitting ending of the country's island status was routine. For good measure he added that the Germans' first Zeppelin flight over London had made the heart of Empire vulnerable, and had seen to it that the imperial capital ceased to be the most secure of the world's great cities. Hoare reviewed developments in long-distance Empire flying and told his listeners about the aim to make Britain as great in the air as on land and sea. His presentation drew a parallel between critics of Empire air services and the reactionaries who 100 years previously had thought that railways and steamers were a gigantic folly and the work of the devil.

Hoare's talk was aired just after Alan Cobham had departed on his first African flight. A year later the BBC broadcast Cobham's return to London. In February 1927, there followed a twenty-minute feature about the Hoares' flight to India and back. In her scripted talk, Lady Maud's most notable remark was about women's role in aviation being identical to men's, namely using flight to knit together and civilise the world. Beyond gender, she saw no need to say who the torchbearers were.

Sir Samuel touched on other romantic themes. His longer address told of their flight as an episode worthy of the *Arabian Nights*, and as a start in making the air 'a new King's highway'. Using the ancient and curious customs of foreigners to signal British elevation, Hoare spoke of the vast number of picturesque spectators their flight attracted. There were wild-looking men and women from black tents in the Libyan Desert, Persian women wearing curious skull-like masks over their faces and Indian women in purdah carts. Best of all, British people *en masse* flocked and cheered. India declared a special holiday. For the British community overseas, *Imperial*'s aircraft epitomised quicker imperial intercourse. Air connections meant letters from home in half the time and a week added to each end of leave. British people welcomed a machine that would bring them nearer to their friends and relatives, and they applauded an achievement of British workmanship and enterprise. One English woman told Lady Maude that she had travelled half across India to see the arrival of the first aeroplane from London. Businessmen welcomed something showing that the spirit of British adventure was alive and that the nerve of British pioneers was not weakening.[67]

At the end of August 1929 the Duchess of Bedford made a radio broadcast about her Indian flight. The following March the BBC broadcast Lt-Cmdr Kenworthy's talk on his return flight to India and, as was customary, published the text in its *Listener* periodical. In 1932, Amy Johnson – by then Mrs Mollison – broadcast from Cape Town on the BBC, explaining that her flight there had partly been 'to keep up British prestige in the air'.[68]

In May 1934, listeners in Britain could hear the National Service broadcast of 'Airways of Empire'. The BBC's *Radio Times* billed the programme as a sound-panorama of the development of flight since the Middle Ages. Whether in the air or on the air it was easy to give flight a parochial slant. Advance notices referred explicitly and exclusively to *Imperial*. Only Britons were interviewed, and those who had aviation affiliations were all associated with that airline. Dreamy wording in the written advertisement for the broadcast told of British aeroplanes reaching out further and further, passing each other in the air 'above all the habitable globe'. The text was wedged between two illustrations. For those in the know, the G-registration visible on the airliner in the clouds at the top of the page denoted that it was British. At the foot of the page the shadow of the plane fell on a camel caravan leaving an oasis.

The publicity may have been more persuasive than the broadcast. The *Star* newspaper, not itself a beacon of taste, condemned the performance as 'rubbish, dull, unintelligible, silly in parts, with idiotic conversation between a half-wit and two barmaids'. Speech was considerably more elevated in the BBC's Empire broadcast of the ceremony at Croydon in December 1934 to mark the start of the regular air mail service to Australia. There was no trumpet fanfare, but the airwaves transmitted speeches by the then Secretary of State for Air (Lord Londonderry) and the Postmaster-General, and the heraldic command which Geddes gave to the flying boat captain: 'Pilot, the King's Mail for Australia!'[69]

Empire aviation featured in two of the seven, twenty-minute weekly pre-recorded talks that the BBC broadcast in 1936 on both its National and Empire services under the title 'Conquest of the Air'. The series was advertised in the BBC's own paper[70] and in *Aeroplane* and *Flight*. Submitted scripts were vetted by the head of the press section at the Air Ministry, where a consultant also made contact with people who could advise on matters such as the risks that tropical diseases created for public health after tropical flights. The series was prefaced by a preliminary feature that set out the highlights of aviation using the words and speech of twelve prominent aviators, interspersed with quotations from *The Times*. Snatches of symphonic music by the Swedish composer Kurt Atterberg set a modern tone. The final programme in the series (compiled by C. W. A. Scott, winner of the 1934 MacRobertson race) featured aeronautical experts from France, Germany, Russia, the United States and Britain. In his otherwise unremarkable contribution to the concluding episode, Sir Francis Shelmerdine, the Director-General of Civil Aviation, made a geographically wayward but propagandistically useful reference to Britain being at the centre of the 'civilising worldwide web' of British air links.

The producer of 'Conquest of the Air' intended to include a talk from Geddes or Woods Humphery on problems of Empire civil air routes. She listed international liaison, currency difficulties, insurance, pilotage problems, passenger comfort, the position and design of aerodromes, opening of new routes, night flying and costs. A follow-up talk by Gen. Groves (at the Air League) was to tackle the question of how Britain's position in the world influenced aviation policy, and how matters stood in relation to the United States, Germany, the USSR, France, Italy and Japan. Both Geddes and Woods Humphery proved uncooperative, and an illness ruled out Groves. In their place, Frederick Montague, an M.P. and Under-Secretary of State for Air in the 1929–31 Labour Government, spoke about 'Airways of the Empire'. An ex-air mail adviser to the Post Office gave a talk on 'Great Britain's Place in World Aviation'. Montague's script recollected the Herculean efforts to make the African airway in 1919. He rehearsed the tale of a project involving courageous exploration and risky adventure in jungles and swamps, in the tracks of great beasts, and among little-known natives belonging to races not always friendly. A new generation learnt the old narrative; broadcast remade legend.

The series of radio talks was judged a success and in the BBC there was talk of making a potted version for the Children's Hour programme. The producer's internally circulated post mortem on the series identified two speakers and consultants as outstanding. Neither of them was involved in the programmes about civil or imperial aviation. On radio, Empire aviation lacked a persuasive spokesperson. The producer described C. G. Grey as 'dangerous'. Shelmerdine was 'a dull speaker, nervous, and useless as a consultant'; Sempill was 'flashy'; Chamier, the Secretary-General of the Air League, was 'intelligent and honest' but a disappointing broadcaster.[71] For its next feature on Empire and aviation the BBC tuned in to other voices.

Empire air transport reappeared on the National Service in 1938 as part of a series of four radio programmes entitled 'Lines on the Map'. The package was about 'selling the Empire to British listeners'. Taking a rather less clinical approach than had been the case in the making of 'Conquest of the Air', the programme makers (including the young poet, novelist and dramatist, John Pudney) devised a dramatic exposition of a flight from England to Australia. They aimed to interview a flying boat captain, a flight steward, a cleaner at Alexandria, an Indian coxswain who worked on a motor launch in India and a mechanic at Brisbane. Pudney invited the young English composer, Benjamin Britten, to write music for the flight sequence; he would have scored a distinctively English sound.

The technical difficulty of timing overseas recordings to coincide with erratic aircraft movements was eventually overcome. And there is no evidence of any comeback about unwarranted advertising: the BBC was sensitive about any 'undue boosting' of *Imperial*. The airline's Press Manager had a contradictory concern. Robert Brenard balked at interviews with subalterns as he did not wish the impression to be given that his Company employed 'cheap native labour'. He need not have been. The issues which new voices were asked to address, unpaid, were innocuous matters such as the climate and the times of arrival and departure of planes. The flying boat steward would be quizzed about the nature and source of the food he served. In an aside, the programme maker conceded that this line of questioning might be dull: 'I rather fear that *Imperial Airways* are the kind of folk who carry everything in tins.' If candid conversation was anticipated, it could be silenced easily. In Brisbane, Qantas Empire Airways, acting for *Imperial*, felt that a talk by a traffic clerk would be more suitable than one by a mechanic. The clerk's recorded talk about aviation was not what the programme producer wanted, but at least the twang in his voice was authentic.[72]

Notes

1 G. H. Pirie, 'Cinema and British imperial civil aviation, 1919–1939', *Historical Journal of Film, Radio and Television*, 23 (2003), 117–31.
2 SAD, *image no. SAD.778/1/72.*
3 Courtenay, *Airman*; *Air Review* (May 1936).
4 G. E. Woods Humphery, 'Imperial airways', *Air Annual of the British Empire*, 5 (1933/34), 7–19.
5 P. B. Ellis and J. Schofield, *By Jove, Biggles! The Life Story of Captain W. E. Johns* (Watford, 2003), p. 111.
6 K. Castle, *Britannia's Children: Reading Colonialism through Children's Books and Magazines (Manchester, 1996).*
7 Ellis and Schofield, *By Jove, Biggles!*, pp. 231–2.
8 W. Courtenay, 'To Calcutta and back', *Flying* (9 July 1938), pp. 6, 7, 29, 30; V. Burnett, 'By flying boat to Africa', *Flying* (17 September 1938), pp. 6, 7, 11.
9 *Popular Flying* (February 1936), pp. 586–9, 624, 626.
10 *Popular Flying* (October 1936), p. 366. Warner would become the well-known naval historian; his employer, Chatto and Windus, had just published Huxley.
11 *Popular Flying* (July 1936), pp. 179–80.
12 *Air Travel and Commercial Air Transport* (January 1934), p. 39; (April 1934), p. 161.
13 James, 'Charles Grey and his pungent pen'; W. Boddy, 'C.G.G.' *Blackwood's Magazine* (1974), 64–76; *The Times* (24 August 1934).
14 *Aeroplane* (20 June 1934; 9 January 1935; 2 December 1936; 28 September, 7 December 1938).
15 C. S. V. I. Stewart, *Airways of the Empire: their History and Development* (University of Nottingham, 1936); *Bystander* (1 May 1934), pp. 224, 226, 228; *Straits Times* (Singapore) (6 March 1934).
16 *Airways* (November 1926; May, June, October, December 1927; March, October 1928; March 1929; September 1930; January 1931).
17 H. L. Crosthwait, 'Air survey and Empire development', *Journal of the Royal Society of Arts*, 77 (1928), 162–81; 'Aerial survey of East and Central African territories', *Journal of the African Society*, 29 (1930), 333–42; *Nature* (22 December 1928), pp. 949–50.
18 M. Cross, 'Aerial photography and Empire development', *World Today*, 56 (1930), 210–17; G. T. McCaw, 'Air survey as a means of empire development', *Crown Colonist*, 3 (1933), 419–22; J. S. Mills, 'Surveying from the air', *United Empire*, 20 (1929), 526–30.
19 *The Times* (8 February 1930); P. R. Burchall, 'An investigation of the possibilities attaching to aerial co-operation with survey, map-making and exploring expeditions', *Journal of the Royal United Services Institute*, 67 (1922), 112–27; Masefield, *To Ride the Storm*, p. 180; W. J. C. K. Cochran-Patrick, 'Air surveys in Burma', *Journal of the Royal Aeronautical Society*, 29 (1925), 603–24; and 'Aerial reconnaissance mapping in Northern Rhodesia', *Geographical Review*, 21 (1931), 213–28; *The Times* (4, 27 January 1930); 'Aerial survey work', *African Air Travel* (August 1932).
20 *Air Review* (September 1934).
21 *The Times* (3 June 1933), p. 9; J. I. Stewart, *An Economic Geography of the British Empire Overseas* (London, 1933).
22 C. B. Fawcett, *A Political Geography of the British Empire* (London, 1933), p. 68.
23 *Field* (7 November 1931); (24 February 1934), p. 400.
24 W. F. Forbes-Sempill, *The Air and the Plain Man* (London, 1931), p. 102; *Field* (19 May 1933), p. 1209; (24 February 1934); (2 June 1934), p. 1289.
25 *Field* (5 November 1932); (25 February, 18 March, 12 August, 30 September, 7 October 1933).
26 *Times Literary Supplement* (23 November 1933), p. 828.
27 G. M. Rogers, 'The stolen idols', in H. Knowlton (ed.), *Empire Annual for Boys* (London, 1933), pp. 136–49.

28 H. D. M. McCutcheon, 'London to Bombay: a thrilling story of an airmail flight and fight', in J. Northcroft (ed.), *Boy's Own Annual* (London, 1934), pp. 635–8.
29 M. Cadogan, *Women with Wings* (London, 1992), pp. 137, 153–4.
30 H. Harper, *The Romance of a Modern Airway* (London, 1930).
31 Salt, *Imperial*, pp. xi–xii; *Times Literary Supplement* (18 December 1930), p. 1075.
32 Cobham, *Australia*; Cobham, 'My flight'; Cobham, *Twenty*.
33 Samson, *Flight*; R. Humble, *Cape Town to Clyde* (London, 1932).
34 B. Bennett, *Down Africa's Skyways* (London, 1932).
35 *Times Literary Supplement* (22 September 1932), p. 658.
36 H. Harper and R. Brenard, *The Romance of the Flying Mail: A Pageant of Progress* (London, 1933); *Aeroplane* (22 November 1933), p. 876.
37 On the return flight Sir Harry photographed Lady Bandon, wife of Earl 'Paddy' Bandon (twenty-nine). Both were flying on *Imperial* to his RAF Cairo posting after their Nairobi wedding. *Bystander* (24 May 1933), pp. 356, 358.
38 H. E. Brittain, *Pilgrims and Pioneers* (London, 1945); Brittain, *By Air*; Brittain, *Wings*; *Times Literary Supplement* (28 December 1933), p. 920; LSE, Brittain 8, 12, 17.
39 *Spectator* (29 December 1933), pp. 971–2; *Times Literary Supplement* (24 January 1935), p. 41.
40 Brittain, *Wings*, pp. 95–7.
41 G. G. Jackson, *Twenty-Six Flying Stories* (London, 1933).
42 G. P. Olley, *A Million Miles in the Air* (London, 1934); *Airways and Airports* (December 1934), p. 349.
43 T. C. Bridges and H. H. Tiltman, *Epic Tales of Modern Adventure* (London, 1935).
44 W. E. Johns (ed.), *Thrilling Flights* (London, 1935).
45 C. S. J. Sprigg, *British Airways* (London, 1934), p. 11; C. S. J. Sprigg, *Great Flights* (London, 1935); A. Williams, *Conquering the Air* (London, 1933).
46 R. Brenard, 'Spanning the Empire by air', in H. Golding (ed.), *The Wonder Book of Aircraft* (London, 1934), pp. 113, 117.
47 *Times Literary Supplement* (30 November 1935; 21 November 1936; 6 November 1937).
48 D. Butts, 'Imperialists of the air-flying stories, 1900–1950', in J. Richards (ed.), *Imperialism and Juvenile Literature* (Manchester, 1989), pp. 126–43.
49 A. O. Pollard, *Romantic Stories of Air Heroes* (London, 1937), pp. 156–7; A. O. Pollard, *The Boy's Romance of Aviation* (London, 1939).
50 J. F. C. Westerman, *The Air Record Breakers* (London, 1937).
51 Butts, 'Imperialists', p. 138.
52 W. E. Johns, *Biggles in Africa* (London, 1936). A similar search in 'cannibal territory' appeared in the *Boy's Own Annual* in 1914. Castle, *Britannia's Children*, p. 103.
53 *Air Review* (April 1936), pp. 34–5.
54 N. MacMillan, *The Romance of Flight* (London, 1935); N. MacMillan, *Sir Sefton Brancker* (London, 1935); A. Johnson, *Sky Roads of the World* (London, 1939).
55 Dower, *Amateur*.
56 *Spectator* (31 August 1934), p. 300; *Times Literacy Supplement* (30 August 1934), p. 590.
57 F. V. Monk and H. T. Winter, *Advance in the Air* (London, 1936); F. V. Monk and H. T. Winter, *Air Mail* (London, 1936). Finch, *World's*.
58 *Spectator* (23 April 1937), p. 729.
59 On the aristocratic writer-pilot's empowering inscription and dis-empowering appropriation of landscapes, places and people see B. Daley, 'Writing from above: representations of landscapes, places and people in the works of Antoine de Saint-Exupéry', *Journal of Cultural Geography*, 26 (2009), 127–47.
60 *Times Literary Supplement* (31 August 1933), p. 569; (24 January 1935), p. 41; (12 November 1938), p. 729; *Spectator* (13 November 1936), p. 869.
61 Coke, *Passenger*. The book also describes crossing the South and North Atlantic aboard *Graf Zeppelin* (1931) and *Hindenburg* (1936).
62 *Spectator* (7 June 1930), p. 934.

63 *Aeroplane* (4 April 1934); Makin, *Swinging*, p. 172.
64 J. Squire, Preface, in R. de la Bère (ed.), *Icarus* (London, 1938).
65 B. Hershey, *The Air Future: a Primer of Aeropolitics* (New York, 1943).
66 There are no imperial references in anthologies of flying poetry published in England in 1925 and 1938. S. W. Murray, *The Poetry of Flight* (London, 1925); de la Bère, Icarus.
67 CU, Templewood, V (5).
68 Russell, *Flying*; *Listener* (5 March 1930), pp. 399–400; *The Times* (19 November 1932), p. 11.
69 Cluett, et al. (eds), *Croydon*, pp. 44–5; *Post Office Magazine* (January 1935), p. 18; *BBC Programme Records, 1934.*
70 *Listener* (18 March 1936), pp. 529–32; (11 March 1936), pp. 469–72; (22 April 1936), p. 782.
71 WAC, R19/64/1 & 2; *BBC Annual, 1936.*
72 WAC, R19/304.

10

Re-flying Empire

The first life of Empire aviation ended in 1939. The loose arrangement of Commonwealth air services which succeeded it after the Second World War was driven by similar impulses to serve Britain's world interests. Within a decade or so, however, vastly superior aeronautical technology all but obliterated Empire air tracks and altered flying experiences significantly. Decolonisation created new international political and economic formations and loosened sentimental ties.

The actuality of imperial flying never quite reached the pitch that was hoped, and a great deal of the service anticipated from Empire aviation in the inter-war period was to reside forever in inter-war imaginings. An imperial imaginary propelled Empire aviation into existence and sustained its development. In the past six decades Empire aviation has been actively re-imagined and reincarnated as historical subject, hobby, and period artefact and icon. The original levitation of Empire, its expression and representation between two world wars, and the projections of imperialism sought from it in the past, have been reconstructed.

The historiography of Empire aviation is one source of imperial flying nostalgia. Its residual magnetism is contained in books, academic papers, newspapers, magazines, biographies, diaries, photographs, maps and memorial lectures. Airfields, buildings, statues, plaques, photographs, paintings, murals, street names and museum exhibits are spaces of fragmentary nostalgia containing, mostly, reminders of machines and the men who flew them. Sound and image reproductions complement these static mementos of flying and help to convey the sensibilities and sensuality of imperial flying.[1]

Memory and nostalgia have gone full circle: dreams of a glorious future for an airborne Empire have been replaced by fantasies of its glorious past. Woven deeply into the fabric of aviation, and into British public memory, the culture of imperial flying is renewed through collections of memorabilia. Prominent among these are reproduction *Imperial Airways* advertising posters, first-flight covers traded in aero philatelic circles, and collectables such as cigarette cards, postcards, cigarette boxes and aircraft kits. All evoke the lost 'golden age' of imperial flying.

Recollection

After Empire flying ceased, the mass media continued as a medium of sporadic and partial recall of the thick confection of Empire aviation. Shortly after the Second World War, the BBC broadcast an interview with Sir Arthur W. Brown. Later, in 1954, on the occasion of the unveiling of a memorial statue at Heathrow, the rigger who accompanied him and Alcock across the Atlantic in 1919 spoke briefly on radio. In the 1950s and 1960s, BBC recordings were made with Sir Alan Cobham and with Capt. O. P. Jones of *Imperial*. Interviews with Amy Johnson were replayed. Imperial flying nostalgia was being stoked.

In June 1961 the Broadcasting Corporation of Rhodesia and Nyasaland produced a programme entitled 'Flight South'. The radio transcript told the story of travel between England and southern Africa as remembered (and spoken) by Capt. O. P. Jones, *Imperial*'s air mail adviser, and three aviation pioneers. Both van Ryneveld and Brand were interviewed. Cobham's anecdotes recalled improvisation and struggle. He remembered once instructing missionaries to test the suitability of ground for landing by running a car over it at 30 m.p.h. He recollected the difficulty of arranging the delivery of aircraft fuel. Cobham remembered fondly the spirit of enterprise and opportunity in colonial Africa. It was typified by the District Commissioner at Abercorn who wanted him to buy land and settle among eighty expatriate coffee farmers: Sir Alan was 'just the type he wanted'.[2]

Intriguing personal recollections of imperial air travel emerged in the BBC's 1979 television series that drew on one social history text and generated another. An ex-*Imperial* manager in Egypt spoke of being expected to know the going rate for Cairo's 'ladies of pleasure'. He and another interviewee told how the airline bus called at Alexandria's nightclubs and brothels on its way to the airport in the early morning, and how one Madame customarily cashed in piles of *Imperial Airways* passenger coupons. A pilot recalled getting into trouble from a Governor's wife when he chose to seat the [prettier] wife of a public works department official on his right at dinner during an overnight stop at Sharjah. His memory of passengers on the India run was of a pompous lot. He found South Africans uncouth, Rhodesians provincial, Kenyans fun and charming. An easy-going nature would have stood them in good stead for the frequent misdirection of luggage 'due to the porters being illiterate'.[3]

Aired in November 1988 on BBC Radio Four, a thirty-minute documentary entitled 'Flights of Fancy' endorsed the view that respectable people showed exotic tastes in entertainment when away from Britain. Another *Imperial* employee recollected independently how passengers staying overnight in Cairo sought out Egyptian belly dancing and cinemas screening pornography. In 1994 the BBC World Service did a radio broadcast on the occasion of what would have been the seventieth anniversary of *Imperial*.[4] A three-part radio series followed in 1999.[5] This time the broadcast was based on the narrative of a modern civilian airliner journey which tracked the original Empire civil air route across fifty-three sectors between England and Australia. Private flying across the Empire was recalled in a 2008 interview on the BBC Woman's Hour programme in association with publication of a book about the Duchess of Bedford.[6]

The airline journey which the BBC revisited in 1999 took as its source Alexander Frater's 1986 book *Beyond the Blue Horizon*. Reminiscences and air-lore throng the pages. Frater's curiosity began as a youth when he presumed *Imperial Airways* was some sort of department of the royal household. His adult safari breathed new life into the mythology and reality of Empire flying. He met old-timers at outstations. He visited historic airfields and hotels written about by Cobham and Hoare, and by passengers and airline employees in their submissions to the *Imperial Airways Gazette*.

By juxtaposing past and present flight, Frater's narrative accentuates the atmosphere of imperial flying. He notes, for instance, that the Handley Page airliners 'majestically emphasised Britain's own somewhat immodest perception of itself', and that pilots were 'public figures, notables in their own right, with loyal bands of fans and followers'. Frater recollects how the flying boats, complete with their maritime fittings, were built by shipwrights who launched them ceremonially down tallow-smeared slipways. A brass bell was rung at an Empire outstation to signal aircraft movements. He reports the celebration of imperial aviation feats in some of Darwin's street names: Vimy, Amy Johnson and Hinkler. The Smith brothers and their mechanics are commemorated similarly.

Anecdotes recovered from written sources, or from long memories, pepper Frater's writing. His own conversations revealed that a Little Englander in a Malaysian post office declined KLM's offer to fly delayed mail to London in 1931 because that airline was not British. Frater discovered that in the heat of a Middle East summer, aircraft mechanics carried their tools in buckets of water. He was told that Chetwynd's career-ending view about the undesirability of flying with Indians was common among non-Indians. Even the first airhostesses hired by *Air India* were all English girls. Racism appeared in the sky in other guises. A resident of Alexandria recalled *Imperial Airways* stewards being patronising, and snobby passengers treating him as a mere 'Gyppo' and a 'wog'. Inverting the common Western image of Orientals as dissolute, Frater humorously presents the sheiks' obstruction on the Arabian air route as their attempt to keep at arm's distance any British civilian aircraft containing 'an unsupervised rabble of gin-sodden infidels'.[7]

An imbiber or not, Michael Powell has written from experience about flying imperially. In his 1986 autobiography, the British film director revealed that he persuaded his boss, Alexander Korda, to send him to Burma to fetch film costumes and props, and make contacts. At the age of eighty-one he recalled his *Imperial* journey from Southampton to Rangoon in 1937 as one of his happiest travel memories. He exaggerated flying on average 2,000 miles a day (half that is more likely), but he tells reliably of the flying boat with mahogany rails, brass fittings and a splendid bow wave.

The jolly Jules Verne adventure filled Powell's head with memories. Flying part of the way in the company of five American oil engineers who disembarked at Bahrain, he landed on isolated, romantic and beautiful water bodies, including private lakes. Black-, brown- and yellow-skinned locals, and one or two enthusiastic and respectful punkah wallahs, dashed, dived and swam to meet the aircraft. At a new refuelling stop used instead of Gwadar, turbaned 'tribesmen' in loincloths climbed on the aircraft floats, took charge of mooring, pumped petrol and poured cans. Powell recalled a moustached *Imperial* 'skipper' who stuffed his potbelly with breakfast at Basra. He encountered two huge, sweaty, silent young aircraft engineers flying a sector on transfer. Sir John Ward, the British Resident at Basra, seemed to Powell to be 'the uncrowned King of Iraq', the local hotel his latest achievement. In memorable events *en route*, two *Imperial* flying boats collided on the water at Sharjah (a replacement was sent from Karachi), Indian villagers shook their fists as the aircraft made buffalo stampede, and passengers fed ham sandwiches to crocodiles in Udaipur's sacred lake.[8]

Inter-war Empire aviation has also been the subject of purely fictional treatment. Here, flight is a scene-setting device rather than the subject. *Imperial 109* (1978) is a thriller set aboard an *Imperial* flying boat on a journey from Durban to New York via London in 1939. The safe delivery of forty-four crates of gold bullion underpins personal tensions that erupt as the flying crew battle the elements. The cabin passengers battle each other in a tale of love, dalliances, intrigue, hate, revenge, murder and heroism. The 'Caterina' flying boat is imaginary but the setting is authentic. The author begins with a nostalgic description of flying boats spanning the immense distances separating the dominions. Standards on board echoed those of leisured and spacious sea travel 'as if in defiance of the new order that was already overtaking the world and would soon sweep away the last trace of the Imperial past'.[9]

Loss of innocence is one of the threads of a racy family saga written by Graham Masterton, author of multiple blockbuster novels. Funding, building and piloting a seaplane for a prestigious air speed race launches *Lords of the Air* (1988). In contrast to social scheming and sexual infidelity, imperial themes are muted, but Capt. Jones, Amy Johnson, Woods Humphery, Lady Houston, Harry Harper and *Imperial Airways* make brief appearances. Social style, personal cunning and a stubborn conservatism are key elements of the novel. Balancing commercial practicality against prestige emerges as a pivotal consideration for aircraft manufacturers.[10]

Binding's 1993 novel *In the Kingdom of Air* harks back to the so-called 'golden age' of British imperial aviation, and recovers one of its moods. The author himself never knew the Empire airline and draws instead from a well of public memory. In a wistful third-person eulogy he recalls Britain's first long-distance overseas airline and its prime – and ultimate – signifiers, the Empire flying boats. Introspective passages conjure a fine image of their gala performance 'in the age of the white saxophone . . . in the dress circle'.[11] Loaded with the Empire's mail and the Empire's emissaries, they flew over everything and were watched by everyone.

Rather like the flagship supersonic 'Concorde' aircraft fifty years later, the flying boats had an unreal aura. Binding has been taken in by their reputation. As Maj. Mayo's scorching 1939 assessment implied,[12] their success originated in publicity rather than practice. Operator, manufacturer and media colluded in a hugely effective promotional campaign to convince themselves and the public about the prospective rewards from a new generation of aircraft. Text drummed up enthusiasm about unrealistic standards of service; lavish images presented a vision of a luxurious and streamlined air Empire anchored in the glorious maritime past. The success of the propaganda can be measured not least by the enduring impression that flying boats generally marked the 'golden age' of British civil aviation. Some of the delusion originates with the way pre-war Empire flying boats are mistily elided with their superior post-war successors. In turn, the latter were romanticised as fast as they were replaced (by 1950) with more utilitarian landplanes on intercontinental Commonwealth routes. The surviving pre-war posters and publicity documents dramatise wishes not reality; they made the

Empire flying boat a false and premature icon. They only partially memorialise a truncated Britannic flirtation with sky and sea.

Empire aviation was about more than overseas air routes and flights. Uniquely, Binding's novel highlights its domesticity. Imperialism, indeed, and Empire flying, had reverberations in Britain. Overseas activities could have an intensely local resonance, as the inhabitants of Rochester knew. There, Short's aircraft factory gave lives focus and rhythm. Schoolchildren stood at their mothers' sides watching trial flights; they charged out of the school gates on a corporation afternoon off. The flying boats were a sign of achievement and prosperity: sheds were specially constructed, new men were hired, new skills taught. In an aircraft town too, people lived in a world of wide horizons. Building and testing an aeroplane produced fresh heroes: the test pilots and the constructors on high ladders sparked boyish imaginations. The factory gave the town common cause and celebration, best marked, perhaps, after the successful return of an aeroplane from its first flight. Taut emotions uncoiled in a pandemonium of public joy in the streets. The townsfolk may have celebrated in 'The Canopus', a local inn that the October 1938 edition of the *Imperial Airways Gazette* reported as being renamed after the first Empire flying boat.

After 1945, the beautiful and beguiling flying boats became anachronisms. They were the wrong shape and they flew at the wrong speed, reflects Binding's protagonist. They even landed on the wrong substance. Despite holding the world in their grasp, their time had passed. Culling began. The planes were held without reprieve at their bases. One by one they were seized and hauled to the shore. Some were crated up and lost, a forgotten jigsaw on a forgotten wharf. Some were left to sink slowly in shallow, unknown waters. Some were broken up. After their wings were clipped, their propeller blades were stacked and hawked on bargain pavements, and their tubular seats were installed ungraciously and anonymously in wayside cafes.

Paralysing the flying boats damaged Rochester. The town at the mouth of the Thames was never the same. Its residents lost more than jobs: 'empty skies led to empty hearts'. People still danced, went to the cinema and pub, and kissed in doorways, but the dismantled flying boats overshadowed these joys. Seen from the main street, the top of a bus and the window of a train, the hulls strung out on the river were an 'unhappy reminder of what we had been and what we had become'. Only the father of Binding's protagonist detaches himself from the romance. He chides his son for his childish interest. Using derogatory colonial language, he mutters that the flying boats were only fit for making the 'Abboes hop around' in Australia and for allowing Englishmen 'to strut among the kaffirs' in Durban.

Binding's tale of passions unfolds against this unhappy backdrop of change on the banks of the Medway. The stranded flying boats become metaphors for loss and freedom. Children could only play at being pilots, stewards and passengers. One obsessive boy swots up technical information about the flying boats and spends hours gazing at the hulks through his bedroom telescope; it is as if the abandoned, de-masted ships – floating prison dungeons – on the Dickensian marshes of a century earlier were haunting the scene. The boy's prize possession is the 'Empire' cigarette card collection of forty-eight pieces that together form a sectional picture of a flying boat. His compulsive father lives in the past when he helped design flying boat interiors. Among his mementoes is an *Imperial* pilot's cap and a flying boat paperweight. Whereas other factory workers became morose when they lost their jobs, this man became deranged. At the firm's long-service dinner in 1948 there is an erotically charged moment when he fondles, kisses and then dismembers and smashes the table centrepiece, an ice sculpture of *Canopus*. Later he hides himself for five days in the best-preserved flying boat, reliving his fantasies until the police find him and commit him to a sanatorium. A successful workers' conspiracy to cut loose the flying boats and tow one to a hide closes the novel.[13]

A flying boat tale which is not fiction, but might as well be, resurfaced at the turn of the twenty-first century. Attesting to an ongoing popular love affair with Empire flying boats, passages from the book were serialised in BBC Radio 4's 'Book of the Week' slot soon after publication in 2000. Graham Coster's *Corsairville* is about the forced landing and subsequent salvage of *Imperial*'s *Corsair* in 1939. This was the aircraft which, after navigation equipment failed, was mired for almost nine months on the Dungu River in northeast Belgian Congo.[14]

After corresponding with dozens of eager British informants, meeting several and seeing their photographs and film, the travel writer Graham Coster reconstructed the way two rescue teams (the first from Short, the second from *Imperial Airways*) worked to dislodge, repair, re-float and fly out the plane. *Imperial* flew them out to Juba, from where they travelled overland to Faradje. Under instructions from the White men, Black men did most of the physical labour. It was an old picture: local Africans dug the flying boat out of the mud, pumped away water, head-portered sludge, and cut and carried heavy ironwood to build a slipway onto the river bank. Hundreds of Africans hauled up the metal whale with a rope. The Belgian District Commissioner arranged a convict chain gang. The local Belgian gold-mine company, which also repaired the aircraft engines, loaned pumping equipment. After the first disastrous rescue attempt, it was decided to deepen and lengthen the take-off surface by damming the river with a barrage. African hands and muscle built a three-mile network of access roads along which trucks carried twenty tons of wood, 170 tons of stone and mountains of non-porous anthill sand used as a sealant.

The Africans working at the crash site were accommodated in twenty grass huts. The creation of this settlement – which the White salvage crews dubbed 'Corsairville' – was scant compensation for the flooding of several African villages upstream. Nothing is known of their or Corsairville's fate in the last seventy years. And it can only be guessed what happened to the uniform worn by the African who was the personal servant to the rescue team – he sewed it from *Corsair*'s upholstered seats. Porcelain hand basins removed to lighten the beached flying boat made their way to a local mission station. Africans wore discarded rivets, piston rings and split pins as ear-studs, rings and bracelets.

Corsairville corrects the 'extravagant evocations' of previous accounts of the fiasco and heroics on the Dungu, and the fairy-story village. Coster uses his revision – and his own travels along *Imperial*'s Africa route – as a vantage point from which to reflect on the persistent British romance with the 'magnificent folly' of flying boats and Empire. Wrestling with the meaning and mesmerism of the technology, he deduces that flying boats reconnected Britain with the sea, and that their halcyon place in British affections harks back to a lost world of grace, civility and luxury. To his surprise, Coster finds that Africa has no memory of the flying boats. Just as British people nourish the comforting memory of the Empire flying boats, so it appears that the fond legend of Corsairville has been willed into existence. It became the foreign embodiment of re-invented English rural arcadia. Coster detects mythical beneficence behind the longing: from Britain the view was of modern British aircraft descending angelically and 'conferring a lasting legacy of community, communications and shelter on a primitive, isolated corner of Africa'.[15] In some eyes, the equation was simple: flying imperially was civilising.

Encore

The theatricality of late imperial flying was first seized on as a prop in 1939. 'Africa Flight', staged at the Richmond Theatre, had a cast of nine, including a professor, an M.P. and a man bearing the title 'Sir'. Reviewed as a 'pallid melodrama', the play resolved a love triangle in the desert after a crash landing: one of the three was shot after drinking more than his fill on a strictly rationed desert march.[16] Alfred Hitchcock followed with his 1940 film 'Foreign Correspondent'. It featured a full-scale mock-up of an *Imperial Airways* Empire flying boat which, in the film, was downed by a German destroyer while crossing the Atlantic to the USA.

In the 1980s, an icon of Empire aviation was used to brand a popular merchandising incentive and customer loyalty scheme in Britain: a stylised flying boat was the centrepiece in the logo of 'air miles' travel discounting that rewarded high-spending consumers with airfares reduced on a fixed pounds–points–miles exchange scale. The intended connotations were surely enterprise, eminence, value, and national identity and pride. Other resonances of Empire aviation were exploited in a 1990s print advertisement for a dessert wine. A flying boat was shown resting on placid water in the background of a sepia photograph depicting elegance, unhurried adventure and conspicuous consumption. Accompanying text named *Imperial* explicitly, and recollected how the airline set unsurpassed levels of comfort and offered indulgent luxuries. One of those was, of course, the perfect drink to accompany a perfect evening meal. Sixty years had passed since *Imperial* criticised port importers for their vulgarity in stooping to advertising.

A decade on, imperial flying was once again used as a backdrop for theatre. Taking the title of a 1930 book, the play 'Scouts in Bondage' parodied boyish British adventure using *Imperial Airways* as a device. Staged in 2009/10 in the King's Head Theatre, London, the satire was 'a ripping story of bare knees, young pluck and stern endeavour'. Like a *Boy's Own* magazine yarn from the 1930s, the self-caricaturing comedy dealt with the trials of four members of the First Little Poddington Scout Troop. Flying by *Imperial Airways* to India for a jamboree with scouts from all over the globe, the boys crash-land in Afghanistan after being hijacked by a suicide bomber. In this dangerous wilderness, outlaws of the East plot against the White races and threaten the stability of the Raj. Caught up in British intelligence operations, the four brave, honest, jolly, pretty, posh boys defeat the Afghan warlords and Russian spies and make it a country in which it is safe to camp. A massive map of British overseas territories backed the stage set. Reproduced *Imperial Airways* posters helped convey a sense of the period.[17] Flying imperially was evidently still a ready reference for London's theatre-going public, and still entertainment.

Fascination with character drew Cambridge audiences to an amateur dramatic company's staging of a new play about Amy Johnson in 2010.[18] More thought provoking was theatre for which an *Imperial Airways* flying boat was the principal image in the publicity poster. 'Faustus in Africa' is a South African adaptation of Christopher Marlowe's 1604 play about a man selling his soul to the devil in exchange for power and knowledge. It was staged in 1995 in several cities in South Africa, Europe and the USA. More than half a century after the demise of

Figure 14 Faustian flight. Artwork for poster advertising the 1995 play 'Faustus in Africa'.

Empire-class flying boats, the renowned South African artist, William Kentridge, seized on their instantly recognisable shape as a marker of time, technology and travel, and welded these to trickery and temptation. A similar full-colour *Imperial Airways* poster (which may have inspired Kentridge) was titled 'Through Africa in Days Instead of Weeks'. Juxtaposing machine and nature, it pictured a white flying boat in a cloudless sky above an efflorescent palm tree on a pristine beach. Kentridge's composition is similar but its tone is sinister not sunny, suggesting damage more than deliverance. In his version, a rather battered flying boat flies over a scraggly palm tree on a barren landscape. Is it fleeing an impending storm or dragging one in its slipstream? Is it emerging from turbulence that has passed? Icon of twentieth-century British imperial contact, was the flying boat angel or devil?

Off the printed page – and off-stage but still theatrical – one of the most evocative ways in which Empire aviation has been remembered is by re-enacting pioneer flights. Like the original, the re-creation of the first trans-Atlantic crossing in 2005 is not strictly an imperial marker.[19] Neither is the planned flight over Everest in a replica aircraft. The air race between London and New Zealand in October 1953 was a more authentically imperial event. There was little hysteria about the race, but it would have re-created a little *frisson* of international competition. Apparently the handicapping formulae prejudiced the British commercial aircraft (entered by Britain's European airline, not by *Imperial*'s Commonwealth equivalent, BOAC). In an uncanny repeat of the MacRobertson result, *Endeavour* was beaten by a KLM civilian airliner carrying sixty-four passengers and twelve crew.[20]

In 1978, RAF Sqn-Ldr David Cyster commemorated the first solo flight to Australia from England. Leaving in his light plane on 7 February, fifty years to the day after Hinkler's departure, the flight to Darwin took twice as long as Hinkler's sixteen days. Hinkler's pace showed his flying skills, but also the warm welcome 'wherever his atlas was painted red'. In 1978, twelve months of planning were needed to deal with tricky political situations, especially in the Middle East. More deeply bureaucratised perhaps than Hinkler's, the flight did echo familiar elements of unpredictable weather, fuel leaks, fatigue, and dependence on the RAF and on other men whose annual wages were less than the cost of filling the fuel tanks. At journey's end in Brisbane, Cyster posed fittingly in front of Hinkler's portrait in the Hinkler Room of the Royal Queensland Aero Club. Back in England, his flight reminded an ex-governess about how she and the elder of two boys she looked after followed Hinkler's progress enthusiastically and learned a lot of geography. Her first flight had been with Cobham in a local aerial circus.[21]

As if infected by Sir Alan Cobham's restless spirit, in mid 1980 Cyster set out in a modified twenty-seven-year-old light aircraft from Hatfield to Cape Town, re-tracing the route pioneered by Cobham in 1925/26. He took with him his wife and a former RAF colleague. They planned to follow Cobham's route as closely as possible, tracking the Nile and passing through East Africa. The intention was to make a film. Echoing a previous generation of flying, British firms associated themselves with the venture: Dunlop supplied tyres, Lucas serviced carburettors and magnetos, Shell supplied fuel and KLG supplied spark plugs.[22] It is not clear what became of the flight. A commemorative flight in 1980 that certainly passed off well was the re-creation of Amy Johnson's departure from Croydon fifty years previously. Special philatelic covers arrived in Darwin by airline service. A Miss J. Schonburg completed her own flight in a light plane along Johnson's route to Darwin.[23]

In 1994, an Australian and an American (with support crew and Shell fuel) marked the seventy-fifth anniversary of the Smiths' 1919 England–Australia flight. The sumptuous photographic record of the journey in a replica Vimy is seductive but is in danger of romanticising pioneer flying: the fleeting glance leaves an image at odds with the reality of both flights. Text is better than pictures at exposing the hardships and setbacks that made the flight last two weeks longer than its predecessor, but even the written account of the 'odyssey' is an entertaining triumphalism. It is also quaintly nautical, being billed as a tale of a ship that accomplished an epic voyage.

Crossing sixteen independent nations patrolled by air traffic and immigration formalities made the 1994 flight unmistakably contemporary, yet there were echoes of the 1919 venture, not least at stops *en route*. In the Persian Gulf the fliers were greeted by a trumpet fanfare and war chant. Fifteen thousand Bedouin and horsemen (and a prince) swelled the festivities. In Karachi the British Deputy High Commissioner held a 1919-costume party at his residence. Schoolchildren, costumed dancers and traditional musicians met the fliers in Kupang. An emergency landing in a Sumatran rice field was the opportunity for further reinvention. One of the crew organised 100 farmers to help him clear a rough landing strip for the support aircraft; modern White man did more than supervise. Curious locals streamed to see the Vimy. Military police and troops kept the crowds orderly, but not before Malayan names and greetings had been added to those scrawled in Arabic and Hindi in the dust on the Vimy's lower wings. In timeless fashion too, excited spectators had to be restrained when the novelty departed.[24]

There have been two attempts to re-trace the pioneering flights made down the length of Africa in the 1920s. The first, in a sixty-year-old bi-plane, resembled the original casualty-prone expeditions rather better than was intended: the 1996 flight was abandoned after repeated delays in southern Italy because of engine failure which meant that air clearances negotiated for flying through Africa had expired. Three years later, a replica Vimy succeeded in what van Ryneveld and Brand had been unable to do in the original *Silver Queen*, namely reach Cape Town without needing a replacement aircraft.

Eighty years after the first effort, pilots were still motivated by sheer personal challenge. And even without the prospective reward of dominating imperial skies and advertising in Empire markets, commercial sponsorship again facilitated the flying. The 1999 pairing of pilots had the benefit of detailed maps and knowledge acquired about African flying conditions. Weather and landing surfaces were little changed. War – rather than wild men and animals – necessitated a routeing away from the Sudan and Ethiopia. Photographs of landscapes, animals and curious African onlookers evoke scenes from 1919/20. An accident and an improvised repair, and pilot dismay and exhaustion, made for a fairly faithful fifty-eight-day commemorative flight.[25]

Even the aero-sporting challenge of being fastest to and from the Cape endures. Seventy years after Alex Henshaw had set his long-standing 1939 record time for a return solo flight between England and Cape Town, it was broken twice within sixteen months by two experienced airline pilots flying solo. Only the cognoscenti will attach much significance to gruelling long-distance flying that shaves minutes off a three to four-day journey. In the words of the newest record breaker, 'the event was not an adventure in the sense that we imagine when reading an exciting novel. For me it was a lifetime achievement but it is something I never want to repeat. The sky-high moments were few and far between, interspersed mostly by fear and dread (engine failure).'[26] The dissipation of pleasure on long air journeys is not unfamiliar, but centennial consciousness may yet spawn commemorative, trans-continental flights in 2019 and 2020. In a post-imperial era, none are likely to be flag-waving events.

Notes

1 Examples are film on www.youtube.com, www.britishpathe.com; colour 'home movie' clips of a flying boat journey through Africa in 1937 at www.efootage.com; the DVD 'British Civil Aviation – Imperial Airways the Definitive Newsreel History 1924–1939' (accessed June 2011).

2 BBC Catalogue of Recorded Talks and Speeches; RAFM, AC 75/2.

3 Hudson, *Air Travel*; Hudson and Pettifer, *Diamonds*, p. 80.

4 BL, NSA B3499/01; Correspondence with Patrick Gillibrand, 19 February 1998.

5 BBC Radio 4 (27 June, 4, 11 July 1999).

6 BBC Radio 4 (16 July 2008); M. Buxton, *The High-Flying Duchess: Mary du Caurroy Bedford, 1865–1937* (Smeeton Westerby, 2008).

7 Frater, *Beyond*, pp. 67, 111.

8 M. Powell, *A Life in Movies* (London, 1992), p. 280.
9 R. Doyle, *Imperial 109* (London, 1978), p. 11.
10 G. Masterton, *Lords of the Air* (London, 1988).
11 T. Binding, *In the Kingdom of Air* (London, 1993), p. 59.
12 See Pirie, *Air Empire*, pp. 226–30.
13 Binding, *Kingdom of Air*, pp. 59–61, 73.
14 A brief, colourful account appeared in the American news magazine *Time* (29 January 1940).
15 Coster, *Corsairville*, pp. 10–11, 192, 256.
16 *The Times* (28 February 1939), p. 12.
17 *The Times* (24 November 2009); www.scoutsinbondage.co.uk (accessed August 2010).
18 www.bawds.org/amy_wonderful_amy7.htm (accessed December 2010).
19 www.vimy.org/vimyatlantic/index.html (accessed June 2006).
20 *Flight* (16 October 1953), p. 542.
21 *Flight* (20 May 1978), pp. 1548–51; *Flight* (1 April 1978), p. 938.
22 *Flight* (12 July 1980), p. 96.
23 *Cluett*, Croydon.
24 P. McMillan, *The Greatest Flight: Reliving the Aerial Triumph that Changed the World* (Atlanta, 1995); P. McMillan, 'The Vimy flies again', *National Geographic* (1995), 2–43.
25 *The Times* (22 February 1996); P. McBride, 'Queen of the African Sky', *National Geographic* (2000), 36–53.
26 www.henshaw-challenge.com/dnn/default.aspx; http://capechallenge.wordpress.com (accessed September 2010).

11

Conclusion

In the twilight of the British Empire, flying imperially was one last expression of the reach of British overseas ambition and style. The routes flown by the designated imperial airline connected London to major imperial cities, and to lesser colonial towns and even minor bush encampments. Like the several hundred private flyers who tracked these routes in the 1920s and 1930s, airline passengers enjoyed the sensations of speed, independence and discovery. In some cases they saved time. It was even possible to recover time, gazing down on and revisiting in quick succession imperial sites from antiquity. Inland from dominion and colonial coasts, aircraft exposed flyers to other British imperial landscapes and cultures. Aircraft left no permanent traces in air or on water, but their repeated arrivals and passages announced, asserted and reproduced a British presence that was more than just flighty. Imperial flying was an act of simultaneous imperial consumption and production.

By securing administrative, cultural and trade ties in the Empire, flying also had a politico-strategic purpose. Moreover, for *Imperial Airways* passengers, and for private purchasers of British light aircraft, flying imperially was also patriotic. Flying in British-built and British-managed aircraft expressed trust and faith in the mother country's men, machines and organisation. Besides being loyal, it was convenient. Inevitably, British air routes and services connected best those parts of the world where British rule, residence and trade were most prominent. In the 1930s there were no other commercial airlines which served all air routes across the British Empire: money and power had claimed the imperial trunk routes for the London-based airline.

Rest stops and aircraft refuelling points in British territory were places where solo and salaried British pilots, and passive flyers, could be understood easily, and where they would feel welcome and at home. These new overseas spaces were part of British cultural embroidery. So too was the comfortable and intelligible social order in *Imperial* aircraft. High fares meant that most passengers were pedigreed members of high society who shared familiar addresses and social codes. They were educated and successful. They worked in mutually recognisable occupations and professions, and travelled respectably for noble reasons. Pilots spoke English, and dressed the part of skilled, responsible, experienced professionals. Stewards were deferential and indulged passengers in as fine a country-house fashion as was possible in a confined space. Expatriate ground crew spoke English, kept foreign premises in a familiar style, were reliable and kept out of sight until needed. All were presumed to have been trained to the highest possible standards in England. Native labourers at airfields and flying boat bases were orderly and obliging – and intriguing subjects for amateur ethnographic comment and photography. In places where aircraft were strange, the users and guardians of the technology felt superior and entitled. Flying imperially meant flying with attitude.

British imperial flying extended supremacist social practices to yet more spaces in the British imperial net. Vague standards of superior service, safety and speed were advertised boldly. They were referred to avidly in letters, diaries, books, flight reports and exhibitions that helped to invoke and re-constitute the heroism, enterprise and atmosphere of Empire. But these tales and images were caricatures of a technology, airline enterprise and 'upwardly mobile' people. They publicised what was remarkable rather than what was ordinary, dependent, callous and regrettable. They also overstated the safety and interest of flying. Ambiguity, dreariness and risk were seldom acknowledged. Forced landings, crashes and delays brought out the stiff upper lip and positive outlook. The successes of women private pilots were contained in a patriarchal and condescending perspective that acknowledged and celebrated extraordinary individuals rather than admit artificially limited merit and ability.

The caricaturing that elevated British aircraft, organisation, male pilots and passengers was also a process of stereotyping other people. Flying was something that few could afford, and that even fewer learned to do on their own. It lent itself to a culture of superiority and exclusivity. Literally and metaphorically, flying was a way of elevating oneself socially and claiming more than just the high ground. Flying imperially was a technique of even greater transcendence. The pretensions of Empire slotted well into the rarity of expensive long-distance air travel. To fly imperially was to distinguish oneself as modern in outlook and behaviour, time pressed, important and well travelled in ways that steamship and railway passengers weren't, let alone colonised people who used non-motorised transport.

Flying imperially was an experience, an expression and always partly an act of imagination. The dreams began after the First World War. Demobbed airforce pilots sought new victories, frontiers and exhilaration by flying first, furthest, fastest or highest. The line between bold and rash was thin. During and after the Great Depression, adventurous, leisured and disillusioned individuals sought out the distraction of an epic journey. In addition, the business of creating an imperial air passenger market in the 1930s drew heavily on selling novelty, sensation and status. *Imperial Airways'* publicity in particular tapped into and re-created the romance of Empire: flying was presented as an easy and quick way to discover and enjoy its diverse landscapes, wildlife, architecture, people and customs. Images and texts conveyed imperial sensibilities even to people who never flew, and spread a fresh imperial consciousness. People who self-identified as British could still live the Empire.

British overseas aviation in the 1930s is now part of a modern imperial imaginary, not to say myth. Mass air travel has redeemed 'primitive' flight which, albeit slow and unpredictable, is remembered as genteel – 'civilised', indeed. Flying imperially was about a privileged journey to successive interesting encounters. But nostalgia for a past 'golden era' feeds on aeronautical heritage publishing and memorabilia which abstract imperial flying from past reality. Aero philately makes flying dates, duration and routes tangible, and recovers the names and distribution of imperial places and correspondents, but leaves the activity of flying unexamined. Viewing archive film partially fills that gap, albeit the view is of an entirely safe, silent and standardised experience of early air travel, ground services and sights. The exceptional artistry of *Imperial Airways* advertising posters invokes innocent, pampered travel and has made them icons of an age, even though the stereotypical imperial gaze from aloft was more differentiated and fluid than airline publicity suggested. The reverse stare of Africans, Arabs and Indians at the loads and follies of aircraft was never part of marketing Empire air travel for fun or convenience.

Other contemporary references to flying imperially, including the commemoration of key aviation events in the British Empire, selectively re-caricature it and gloss the cultures of imperial flying. Many of the most public residues of imperial civil flying are positive evocations of the people, circulation and tone of late Empire. There are abundant reminders about the privilege and pleasure of commercial flight. There are fewer obvious signs that it was also a wilful performance – a re-staging of Empire. An honest record should also tell of the ambitions, needs, differences and prejudices which flight nourished.

Aviation may not have prolonged Britain's historically maritime Empire, but it did give imperialism new dimensions, meanings and significance. It created new spaces of practice and, in ways that proponents and participants could never have imagined, Empire aviation left behind the kind of sentiment that abrupt endings can generate. Even in parody, British imperial flying survived the end of Empire and became a period icon.

INDEX

Note: page numbers in *italics* refer to illustrations